UNITEXT for Physics

Series Editors

Michele Cini, University of Rome Tor Vergata, Roma, Italy

Stefano Forte, University of Milan, Milan, Italy

Guido Montagna, University of Pavia, Pavia, Italy

Oreste Nicrosini, University of Pavia, Pavia, Italy

Luca Peliti, University of Napoli, Naples, Italy

Alberto Rotondi, Pavia, Italy

Paolo Biscari, Politecnico di Milano, Milan, Italy

Nicola Manini, University of Milan, Milan, Italy

Morten Hjorth-Jensen, University of Oslo, Oslo, Norway

Alessandro De Angelis, Physics and Astronomy, INFN Sezione di Padova, Padova, Italy

UNITEXT for Physics series publishes textbooks in physics and astronomy, characterized by a didactic style and comprehensiveness. The books are addressed to upper-undergraduate and graduate students, but also to scientists and researchers as important resources for their education, knowledge, and teaching.

Jaime Merino · Alfredo Levy Yeyati

Many-Body Techniques in Condensed Matter Physics

Lecture Notes and Exercises for an Introductory Course

 Springer

Jaime Merino
Departamento de Física Teórica de la
Materia Condensada
Universidad Autónoma de Madrid
Madrid, Spain

Alfredo Levy Yeyati
Departamento de Física Teórica de la
Materia Condensada
Universidad Autónoma de Madrid
Madrid, Spain

ISSN 2198-7882   ISSN 2198-7890 (electronic)
UNITEXT for Physics
ISBN 978-3-031-55145-1   ISBN 978-3-031-55143-7 (eBook)
https://doi.org/10.1007/978-3-031-55143-7

Preface

These lecture notes correspond to a postgraduate course on Quantum Field Theory in Condensed Matter that we have been lecturing in the past few years at Universidad Autónoma de Madrid. The main aim of these notes is to introduce the basic concepts and techniques which, in our opinion, are necessary for describing equilibrium and out-of-equilibrium many-body condensed matter systems.

The present notes cover the material corresponding to our 40-hour course which was divided into 30 h theory classes and 10 h exercise classes. While Part I of the book, Chaps. 2–6, introduces equilibrium many-body techniques, Part II, Chaps. 7–9, is devoted to the theory of non-equilibrium quantum many-body systems. These two first parts are based on the more conventional operator formalism. For completeness, we also provide the alternative path integral formulation in Chaps. 10 and 11 in Part III of the book. Finally, the appendix provides hints for the solution of the exercises included in the book. Most exercises involve derivations of key expressions which would be too lengthy to be included in the main text.

Although these lectures are based on many cornerstone books on quantum many-body theory (as cited in the general bibliography) they also reflect our personal views and experience in the subject, as can be appreciated by the selection of problems that are proposed as exercises. In this respect, we have chosen to emphasize the practical aspects of the course, like the application of the formalism to concrete physical situations, rather than the more abstract concepts in the theory. The interested reader should consult the bibliography regularly in order to cover the unavoidable gaps present in our book.

We expect that this material will be useful both for students entering the field as well as for professors teaching a similar crash introductory course on many-body physics in condensed matter.

<table>
<tr><td>Madrid, Spain</td><td style="text-align:right">Jaime Merino</td></tr>
<tr><td>November 2023</td><td style="text-align:right">Alfredo Levy Yeyati</td></tr>
</table>

Acknowledgements

We are grateful to our colleagues in the Theoretical Condensed Matter Dpt.: Profs. Fernando Flores, Alvaro Martín Rodero, and Carmina Monreal with whom we have shared the quantum many-body theory course given at UAM for many years. J. M. acknowledges Prof. Brad Marston for his inspiring Quantum Many-Body Theory course at Brown University. The renormalization group approach introduced in Chap. 11 of the book follows closely that course. Both authors acknowledge specially Prof. Alvaro Martín Rodero for his notes on some aspects of diagrammatic techniques which have been useful to elaborate the present course.

We would also like to thank the students who read through various parts of the text giving us helpful feedback on the text and the resolution of the exercises: Manuel Fernández López, Miguel Alvarado, Francisco Fernández Carbayo, Fátima Cerezo, and Javier Castro.

Finally, we wish to thank our families for their continuous support and patience.

General Bibliography

The present lectures are based on several textbooks on quantum many-body theory. Readers are encouraged to study these textbooks whenever they want to seek for deeper understanding and more details. We provide a list of the relevant books together with the chapters in which they have been mostly used:

[1]   N. Ashcroft and D. Mermin, *Solid State Physics*, Philadelphia, (Saunders College, 1976). (Chap. 1)

[2]   P. Phillips, *Advanced Solid State Physics*, (CRC Press, 2021). (Chap. 1)

[3]   A. A. Abrikosov, L. P. Gorkov, I. E. Dzyaloshinskii, *Methods of quantum field theory in statistical physics*, (Dover, 1975). (Chaps. 2–4)

[4]   A. L. Fetter and J. D. Walecka, *Quantum theory of many-particle systems* (McGraw Hill College, 1971). (Chap. 2)

[5]   G. D. Mahan, *Many-particle physics*, 3rd Edition, (Springer Science + Business Media, LCC, 2000). (Chap. 5)

[6]   P. Coleman, *Introduction to many-body physics*, (Cambridge University Press 2015). (Chaps. 1–5)

[7]   H. Bruus and K. Flensberg, "Many-body quantum theory in condensed matter physics", (Oxford University Press 2017). (Chaps. 1–5)

[8]   R. D. Mattuck, *A guide to Feynman diagrams and the many-body problem*, 2nd Edition, (Dover 1992). (Chap. 6)

[9]   J. Rammer, *Quantum Field Theory of Non-equilibrium States*, (Cambridge University Press, 2007). (Chaps. 7–9)

[10] A. Altland and B. D. Simons, *Condensed Matter Field Theory*, Cambridge University Press (2010). (Chaps. 7–11)

[11] A. Kamenev, *Field theory of non-equilibrium systems*, (Cambridge University Press, 2011). (Chaps. 7–10)

[12] J. W. Negele and H. Orland. Quantum Many-Particle Systems, (Addison-Wesley, 1988). (Chap. 10)

[13] N. Nagaosa. *Quantum Field Theory in Condensed Matter Physics*, (Springer, 2010). (Chaps. 10–11)

[14] A. M. Tsvelik, *Quantum field theory in condensed matter physics*, (Cambridge University Press, 1995). (Chap. 11)

Contents

1   Introduction to Many-Particle Physics in Condensed Matter 1
  1.1   Quantum Field Theory and Statistical Mechanics 2
  1.2   Second Quantization 3
  1.3   Model of Metals 6
  1.4   Hubbard Model 9
  1.5   Heisenberg Model 11
  1.6   Anderson Model of Magnetic Impurities in Metals 13

Part I   Equilibrium Many-Body Techniques

2   Introduction to Green-Function Methods 19
  2.1   Time Evolution 20
  2.2   Time-Ordered or Causal Green's Functions at $T = 0$ 21
  2.3   Causal Green Function for Free Fermions 22
  2.4   Physical Observables 24
  2.5   General Analytic Properties of Green Functions
        in Interacting Systems 26
  2.6   Retarded and Advanced Green Function 28
  2.7   Fermi Liquid Properties 29

3   Perturbation Theory at Zero Temperature 33
  3.1   Interaction Picture 33
  3.2   Time-Evolution Operator 34
  3.3   Relation Between Heisenberg and Interaction Pictures 36
  3.4   Adiabatic Hypothesis 37
  3.5   Gell-Mann and Low Theorem 39
  3.6   Wick's Theorem 40
  3.7   Diagrammatic Approach in Coordinate Space 41
  3.8   Diagrammatic Approach in Momentum Space 46
  3.9   Self-energies and Dyson Equation 48
  3.10 Physical Interpretation of the Self-energy 50

4 **Finite Temperature Green Function Formalism** 55

4.1 Schrödinger Representation 55

4.2 Heisenberg Representation 56

4.3 Imaginary Time Green Function 56

    4.3.1 Matsubara Representation of the Imaginary Time Green Function 58

4.4 Spectral Decomposition of Matsubara Green Function 59

4.5 Perturbation Theory at Finite Temperature 60

    4.5.1 Interaction Representation 61

    4.5.2 Perturbation Theory and Wick's Theorem 61

4.6 Feynman Rules for the Coulomb Interaction at Finite-T 63

4.7 Feynman Diagrams at Finite Temperature 63

    4.7.1 Self-consistent Hartree-Fock Approach 64

4.8 Diagrammatic Approach to the Electron-Phonon Interaction 66

    4.8.1 Electron-Phonon Interaction in Second Quantization 66

    4.8.2 Feynman Diagrams 68

4.9 Evaluation of Matsubara Sums 70

5 **Linear Response and Collective Modes** 75

5.1 Collective Modes in a Fermi Liquid 77

5.2 RPA Approximation 79

5.3 Screening of Coulomb Repulsion 82

5.4 Screening of an External Potential in the Electron Liquid 85

6 **Spontaneous Symmetry Breaking and Mean-Field Theory** 87

6.1 Generalized Green Function Propagator 87

6.2 Application to Metallic Ferromagnetism 89

6.3 Elementary Excitations in a Broken Symmetry Phase 90

Part II Non-equilibrium Many-Body Techniques

7 **Introduction to Non-equilibrium: The Keldysh Contour** 97

8 **Perturbative Expansion in the Non-equilibrium Formalism** 101

8.1 Basic Properties of Non-equilibrium GFs 103

8.2 Perturbative Expansion for a One-Body Perturbation 105

8.3 Case of Many-Body Interactions 107

8.4 The Dyson Equation as an Intregro-Differential Equation 109

8.5 The Triangular Representation 109

8.6 Langreth Rules 111

8.7 Frequency Representation 113

9 **Applications: Electron Transport at the Nanoscale** 115

9.1 Current Fluctuations 121

9.2 Full Counting Statistics 122

9.3 Superconducting Transport 125
 9.3.1 Nambu Formalism 126
 9.3.2 N/S Channel: Andreev Reflection 127
 9.3.3 S/S Channel: Josephson Effect 131
 9.3.4 Transport Through Majorana Nanowires 133

Part III Path Integral Formulation of the Quantum Many-Body Problem

10 Introduction to Path Integral Methods 139
 10.1 Path Integral of a Single Boson 139
  10.1.1 Green Function from Path Integration 143
  10.1.2 Single Boson 144
  10.1.3 Single Fermion 147
  10.1.4 Wick's Theorem and Perturbation Theory 154
 10.2 Path Integral Formulation for Non-equilibrium Systems 157
  10.2.1 The Case of Lattice Models for Electron Transport 163

11 Application of Path Integral Methods: The Renormalization Group Approach 165
 11.1 Non-interacting Fermion Model Under RG: Scale Invariance and Fixed Points 165
 11.2 The One-Dimensional Hubbard Model Under RG: Spin-Charge Separation 170
  11.2.1 RG Perturbative Treatment 172
  11.2.2 First Order Corrections 173
  11.2.3 Second Order Corrections 175
  11.2.4 Spin Sector 178
  11.2.5 Charge Sector 179
  11.2.6 Discussion of RG Analysis of the 1D Hubbard Model 179

Appendix: Hints for Solving Exercises 183

References 213

Chapter 1
Introduction to Many-Particle Physics in Condensed Matter

In condensed matter physics we are interested in understanding the physics of systems with a large number ($N \sim 10^{23}$) of interacting quantum particles. These systems are interesting since they can lead to new collective phenomena displaying emergent properties not present in their individual constituents. Fascinating examples of these phenomena, among others, are superconductivity, superfluidity and the quantum Hall effect. New concepts have risen from the study of such systems such as the concept of quasiparticle, spontaneous symmetry breaking and renormalization common to other fields.

Quantum field theory (QFT) is the quantum mechanics of systems with infinite degrees of freedom. Quantum systems of many interacting particles are difficult to deal with since they imply huge Hilbert spaces which cannot be diagonalized exactly as standard quantum mechanics dictates, even with the largest available computers! QFT is the appropriate framework to explore these systems. QFT in particle physics allows describing the elementary particles (electrons, quarks) and their interactions (photons, gluons) on equal footing. In a similar way, elementary excitations in condensed matter such as phonons in solids or Landau quasiparticles in Fermi liquids are well described as excitations of quantum fields.

QFT allows the exploration of the many-body effects generally neglected in the conventional band theory of solids which has been extremely successful in explaining the differences between metals, insulators and semiconductors. Indeed, many elementary properties of metals can be understood in terms of non-interacting particles. For instance, the temperature dependence of the specific heat of a metal is:

$$C_V \sim \gamma T + A T^3, \tag{1.1}$$

where the first contribution comes from independent electrons and the second contribution comes from independent phonons. At sufficiently low temperatures, the electron contribution dominates.

In spite of the evidence for independent electrons in conventional metals (Al, Au, Ag, Cu), electrons and ions interact strongly among themselves and among each other. Hence, it is really surprising that the one-electron theory for metals

© The Author(s), under exclusive license to Springer Nature Switzerland AG 2024
J. Merino and A. L. Yeyati, *Many-Body Techniques in Condensed Matter Physics*,
UNITEXT for Physics, https://doi.org/10.1007/978-3-031-55143-7_1

actually works in conventional metals! It was the genius of Landau who settled this paradox introducing the final theory of metals known as the Landau theory of the interacting Fermi liquid. In spite of this early success it was soon realized that many physical phenomena observed in solids such as the cohesive energies of metals, superconductivity and magnetism cannot be described just based on one-electron approaches. In fact, it is the interactions between electrons and ions that are ultimately responsible for these phenomena. More exotic and modern physical phenomena such as the fractional quantum Hall effect, heavy fermion behavior, Kondo effect, high-Tc and/or organic superconductivity arise as a consequence of strong electronic interactions. Hence, it is a fundamental challenge in condensed matter physics to understand the effect of electron interactions in these systems. QFT is the most successful framework to achieve this goal.

After pointing out the general connection between QFT and statistical mechanics, we will introduce second quantization as the natural language to deal with many-particle systems. We finally discuss some relevant models in condensed matter expressed in second quantization.

1.1 Quantum Field Theory and Statistical Mechanics

QFT being a statistical theory works with statistically averaged quantum averages. Hence, there is double averaging. In the grand canonical ensemble, this means that the average value of an observable, A, of a system in equilibrium with a heat/particle bath at temperature T is:

$$\langle A \rangle = \frac{1}{Z} \sum_{\lambda, N} e^{-\beta(E_\lambda(N) - \mu N)} \langle \lambda | A | \lambda \rangle = \frac{1}{Z} Tr\{e^{-\beta(H - \mu N)} A\}, \qquad (1.2)$$

where λ are the eigenstates of the system with N particles with a corresponding energy, $E_\lambda(N)$. Note that the trace can be over any many-body basis. For example, the average particle number reads:

$$\langle N \rangle = \frac{1}{Z} Tr\{e^{-\beta(H - \mu N)} N\}. \qquad (1.3)$$

The partition function reads:

$$Z = \sum_{\lambda, N} e^{-\beta(E_\lambda(N) - \mu N)} = Tr\{e^{-\beta(H - \mu N)}\}. \qquad (1.4)$$

In the $T \to 0$ limit this expression reduces to the quantum mechanical expression for $\langle A \rangle$:

$$\langle A \rangle \to \frac{e^{-\beta(E_G(N) - \mu N)}}{e^{-\beta(E_G(N) - \mu N)}} \langle \Psi_G(N) | A | \Psi_G(N) \rangle, \qquad (1.5)$$

where here N refers to the particle sector in which the ground state of the system is found.

From the partition function Z we can obtain all thermodynamic properties. The grand potential reads:

$$\Omega = -K_B T \ln(Z). \tag{1.6}$$

From thermodynamics Ω satisfies the relation:

$$\Omega \equiv E - TS - \mu N. \tag{1.7}$$

which combined with the first and second law of thermodynamics leads to the thermodynamic identity:

$$d\Omega = -SdT - PdV - Nd\mu, \tag{1.8}$$

by which $\Omega = \Omega(T, V, \mu)$. The fundamental thermodynamic relation allows to extract entropy, pressure and the number of particles from its partial derivatives:

$$S = -\frac{\partial \Omega}{\partial T}\Big|_{V,N}; \ P = -\frac{\partial \Omega}{\partial V}\Big|_{T,N}; \ N = -\frac{\partial \Omega}{\partial \mu}\Big|_{T,V}. \tag{1.9}$$

Hence, from the knowledge of Z and the above partial derivatives we can obtain thermodynamic variables of interest.

Generally, it is also interesting to obtain correlation functions analogous to the ones in classical statistical mechanics which allow to characterize the ground state of the system analogous to the ones studied in classical statistical mechanics:

$$\langle n(\mathbf{x}_1)n(\mathbf{x}_2)\rangle \tag{1.10}$$

which gives density-density spatial correlations or time-dependent counterparts:

$$\langle n(\mathbf{x}_1, t_1)n(\mathbf{x}_2, t_2)\rangle \tag{1.11}$$

which also characterize excitations of the system. These are analogous to the Green functions in QFT we will study in this course.

1.2 Second Quantization

Second quantization is a useful language for analyzing interacting many-body systems. For instance, we would like to find the ground state of the many-body hamiltonian above given by: $\Psi(\mathbf{x}_1, \mathbf{x}_2, ..., \mathbf{x}_N)$. Even finding the ground state of the H_2 molecule or He atom which only contain two electrons is a very involved problem. Since identical particles come only in two kinds regarding their properties under exchange. The particles which are eigenstates of the exchange operator with eigenvalue $+1$ (symmetric under exchange) are called bosons and with eigenvalue -1 (antisymmetric under exchange) are called fermions. Imagine we have a set of orthonormal single-particle states: $\{\phi_i(\mathbf{x})\}$. Then we could construct a many-body wave function as:

$$\Psi(\mathbf{x}_1, \mathbf{x}_2, ..., \mathbf{x}_N) = \phi_1(\mathbf{x}_1)\phi_2(\mathbf{x}_2)...\phi_N(\mathbf{x}_N) \tag{1.12}$$

which does not obey the Fermi statistics since the wavefunction should be antisymmetric under the interchange of two identical particles:

$$\Psi(\mathbf{x}_1, \mathbf{x}_2, ..., \mathbf{x}_i, ..., \mathbf{x}_j, ..., \mathbf{x}_N) = -\Psi(\mathbf{x}_1, \mathbf{x}_2, ..., \mathbf{x}_j, ..., \mathbf{x}_i, ..., \mathbf{x}_N).$$

However, the Slater determinant:

$$\Psi(\mathbf{x}_1, \mathbf{x}_2, ..., \mathbf{x}_N) = \frac{1}{\sqrt{N!}}
\begin{vmatrix}
\phi_{i_1}(\mathbf{x}_1) & ... & \phi_{i_1}(\mathbf{x}_N) \\
\cdot & ... & \cdot \\
\cdot & ... & \cdot \\
\cdot & ... & \cdot \\
\phi_{i_N}(\mathbf{x}_1) & ... & \phi_{i_N}(\mathbf{x}_N)
\end{vmatrix} \tag{1.13}$$

satisfy the antisymmetry under exchange for identical fermions. Also note that automatically if $\mathbf{x}_l = \mathbf{x}_m$ or $i_l = i_m$ the determinant is zero satisfying the Pauli principle by which two electrons cannot have the same quantum numbers (including spin).

We can specify the wavefunction $\Psi(\mathbf{x}_1, \mathbf{x}_2, ..., \mathbf{x}_N)$ by just giving the occupation number, $n_l = 0, 1$ for fermions in each one-electron orbital l. In Dirac notation this means that $|n_1, n_2, n_3, ...\rangle$ specifies completely the many-body wavefunction. For example, $|1, 1 >$ is equivalent to the two particle state : $\frac{1}{\sqrt{2}}(\phi_1(\mathbf{x}_1)\phi_2(\mathbf{x}_2) - \phi_1(\mathbf{x}_2)\phi_2(\mathbf{x}_1))$ which is the two particle Slater determinant. The states $|n_1, n_2, ...\rangle$ with all possible values of n_i define the Fock space. The basis vector states can have any number of particles. The vacuum state with no particles is denoted by $|0\rangle$. The occupation number formalism allows to forget the coordinates and orbitals.

We can get practical results with the occupation number formalism in the Fock space by introducing two new quantum mechanical operators acting on the Fock space creating and destroying particles in particular wavefunction. We define the creation and destruction operators as:

$$c_n^\dagger |n_1, n_2, ..., n_n, ...\rangle = (-1)^{n_1+n_2+n_3+n_{n-1}}\sqrt{1 - n_n}|n_1, n_2, n_3, ..., n_n + 1, ...\rangle$$
$$c_n |n_1, n_2, ..., n_n, ...\rangle = (-1)^{n_1+n_2+n_3+n_{n-1}}\sqrt{n_n}|n_1, n_2, n_3, ..., n_n - 1, ...\rangle.$$

The vacuum state satisfies:
$$c_i|0\rangle = \langle 0|c_i^\dagger = 0. \tag{1.14}$$

These operators satisfy anticommutation relations:

$$\{c_i^\dagger, c_j^\dagger\} = \{c_i, c_j\} = 0, \tag{1.15}$$

which imply: $(c_i^\dagger)^2 = (c_i)^2 = 0$. Also we have:

$$\{c_i, c_j^\dagger\} = \delta_{ij}. \tag{1.16}$$

The number operator counting the number of electrons on the state n reads:

$$n_n = c_n^\dagger c_n. \tag{1.17}$$

Using the second quantization formulation, a general many-particle fermionic state can be written compactly as:

$$|n_1, n_2, \ldots\rangle = (c_1^\dagger)^{n_1} (c_2^\dagger)^{n_2} \ldots |0\rangle, \tag{1.18}$$

so we only need the $c_i^\dagger$, c_i operators and the vacuum state $|0\rangle$ to construct a general many-body state in the occupation formalism. For example, $|1, 1\rangle = c_1^\dagger c_2^\dagger |0\rangle$, which is equivalent to the Slater determinant of two particles in the two states, $\phi_1(\mathbf{x})$, $\phi_2(\mathbf{x})$, described before.

Physical observables can be obtained in second quantization. For example, the one-electron potential acting on a state in first quantization and using Dirac notation:

$$V(\mathbf{x})\phi_n(\mathbf{x}) = \langle \mathbf{x}|V(\mathbf{x})|n\rangle = \sum_m \langle \mathbf{x}|m\rangle\langle m|V|n\rangle = \sum_m V_{mn}\phi_m(\mathbf{x}), \tag{1.19}$$

where:

$$V_{mn} = \langle m|V|n\rangle = \int \phi_m^*(\mathbf{x})V(\mathbf{x})\phi_n(\mathbf{x}). \tag{1.20}$$

The above one-electron potential operator can be written in second quantization formalism as:

$$V \equiv \sum_{m,n} c_m^\dagger V_{mn} c_n. \tag{1.21}$$

One can check that this is equivalent to the expression for V in first quantization, since:

$$V\phi_n(\mathbf{x}) = \left(\sum_{m',n'} c_{m'}^\dagger V_{m'n'} c_{n'}\right)c_n^\dagger|0\rangle = \sum_{m'n'} c_{m'}^\dagger V_{m'n'}(\delta_{n'n} - c_n^\dagger c_{n'})|0\rangle$$

$$= \sum_{m'} c_{m'}^\dagger V_{m'n}|0\rangle = \sum_{m'} V_{m'n}|m'\rangle = \sum_{m'}\langle m'|V|n\rangle|m'\rangle = V|n\rangle = V\phi_n(\mathbf{x}).$$

$$\tag{1.22}$$

Alternatively, we can use field operators to obtain expressions in second quantization. Field operators which create a particle at position $\mathbf{x}$ in space: $|\mathbf{x}\rangle = \Psi^\dagger(\mathbf{x})|0\rangle$, can be expanded in terms of a complete orthonormal set of one-electron wavefunctions, $\{\phi_m(\mathbf{x})\}$ of our choice. In the Dirac notation, $\phi_m(\mathbf{x}) = \langle \mathbf{x}|m\rangle$ and using the identity:

$\int d\mathbf{x}|\mathbf{x}\rangle\langle\mathbf{x}| = 1$ then: $|m\rangle = \int d\mathbf{x}\langle\mathbf{x}|m\rangle|\mathbf{x}\rangle$. Using the second quantized form of the basis set: $|m\rangle = c_m^\dagger|0\rangle$ we have:

$$c_m^\dagger = \int d\mathbf{x}\langle\mathbf{x}|m\rangle\Psi^\dagger(\mathbf{x}) = \int d\mathbf{x}\phi_m(\mathbf{x})\Psi^\dagger(\mathbf{x}) \tag{1.23}$$

which can be inverted to give an expression for the field operator expanded in the orthonormal set: $\{\phi_m(\mathbf{r})\}$.

$$\Psi^\dagger(\mathbf{x}) = \sum_m \langle m|\mathbf{r}\rangle c_m^\dagger = \sum_m \phi_m^*(\mathbf{x})c_m^\dagger. \tag{1.24}$$

Hence, the potential energy operator, V, can be expressed in terms of the field operators above:

$$V = \sum_{mn} c_m^\dagger V_{mn}c_n = \sum_{mn}\int d\mathbf{x}\langle\mathbf{x}|m\rangle\Psi^\dagger(\mathbf{x})\int d\mathbf{x}''\langle m|\mathbf{x}''\rangle V(\mathbf{x}'')\langle\mathbf{x}''|n\rangle\int d\mathbf{x}'\langle n|\mathbf{x}'\rangle\Psi(\mathbf{x}')$$

$$= \int d\mathbf{x}\Psi^\dagger(\mathbf{x})V(\mathbf{x})\Psi(\mathbf{x}), \tag{1.25}$$

where the completeness relations: $\sum_m\langle\mathbf{x}|m\rangle\langle m|\mathbf{x}''\rangle = \delta(\mathbf{x}-\mathbf{x}'')$ and $\sum_n\langle\mathbf{x}''|n\rangle\langle n|\mathbf{x}'\rangle = \delta(\mathbf{x}''-\mathbf{x}')$, have been used.

1.3 Model of Metals

We now construct an approximate model for interacting metals applying the second quantization formalism. We start by considering the general Hamiltonian of a solid:

$$H = \sum_I \frac{P_I^2}{2M} + \sum_i \frac{p_i^2}{2m} + \frac{(Ze)^2}{2}\sum_{I\neq J}\frac{1}{|\mathbf{R}_I - \mathbf{R}_J|} + \frac{e^2}{2}\sum_{i\neq j}\frac{1}{|\mathbf{x}_i - \mathbf{x}_j|} - Ze^2\sum_{i,I}\frac{1}{|\mathbf{x}_i - \mathbf{R}_I|} \tag{1.26}$$

where m is the electron charge and M and Ze are the nuclear mass and charge of a lattice of identical atoms, respectively. $\mathbf{r}_i$ and $\mathbf{R}_I$ refer to electron and nucleii coordinates, respectively. Note that we are taking the electron charge $e > 0$.

It is convenient to separate the electron degrees of freedom into conduction electrons localized around the nuclei and the conduction electrons which can propagate throughout the solid. We consider a simplified model for the description of simple metals like the Alkali metals such as Li, Na or K. These consist of closed shell cores and a single valence electron in an s-orbital. For instance, Na atoms with nuclear charge $Z = 11$ have 11 electrons which are arranged in the atomic structure: $1s^2 2s^2 2p^6 3s$. Hence, the single s electron in the $n = 3$ shell sees the effective potential of the nucleus and the core electrons which have an effective positive charge of $Z_{ion} = +1$.

Since the ionic mass is much larger than the electron mass, $M >> m$, the velocity of the electrons is much larger than the ion velocities: $v_e \sim 10^3 v_{ion}$ implying that the ions can be considered, to a first approximation, as being static with respect to the electron motion (Born-Oppenheimer approximation). Hence, an adequate description of a simple metal considers the ions (nucleii+core electrons) fixed at the equilibrium positions of the solid:

$$H_{metal} = \sum_i \left[\frac{p_i^2}{2m} + \sum_I V_{e-ion}(\mathbf{x}_i - \mathbf{R}_I^{eq}) \right] + \frac{e^2}{2} \sum_{i \neq j} \frac{1}{|\mathbf{x}_i - \mathbf{x}_j|} = H_{oe} + V_{ee}.$$

$$(1.27)$$

This is the starting point for describing the dynamics of an interacting electron gas. Due to its importance in the Fermi liquid theory of metals we will analyze this hamiltonian in detail during the present course.

The full hamiltonian for the metal expressed in terms of field operators reads:

$$H_{metal} = \sum \int d\mathbf{x} \Psi^\dagger(\mathbf{x})(-\frac{\hbar \nabla^2}{2m} + V_{e-ion}(\mathbf{x}))\Psi(\mathbf{x}) + V_{ee} \qquad (1.28)$$

where:

$$V_{ee} = \frac{1}{2} \int d\mathbf{x} d\mathbf{x}' \Psi^\dagger(\mathbf{x})\Psi^\dagger(\mathbf{x}') \frac{e^2}{|\mathbf{x} - \mathbf{x}'|} \Psi(\mathbf{x}')\Psi(\mathbf{x}). \qquad (1.29)$$

In principle, we can use any kind of complete orthonormal basis set, $\{\phi_m(\mathbf{x})\}$ to describe the conduction electrons of the metal. Hence, the hamiltonian can be expressed as:

$$H_{metal} = H_{oe} + V_{ee}, \qquad (1.30)$$

where the one-electron part reads:

$$H_{oe} = \sum_{lp} V_{lp} c_l^\dagger c_p, \qquad (1.31)$$

and the electron-electron interaction term is:

$$V_{ee} = \frac{1}{2} \sum_{mnlp} V_{mnpl} c_m^\dagger c_n^\dagger c_l c_p \delta_{\sigma_m \sigma_p} \delta_{\sigma_n \sigma_l}. \qquad (1.32)$$

where the one-particle matrix elements read:

$$V_{lp} = \langle l | - \frac{\hbar^2 \nabla^2}{2m} + V_{e-ion}(\mathbf{x}) | p \rangle = \int d\mathbf{x} \phi_l^*(\mathbf{x})(-\frac{\hbar^2 \nabla^2}{2m} + V_{e-ion}(\mathbf{x}))\phi_p(\mathbf{x}),$$

$$(1.33)$$

and the two-particle matrix elements:

$$V_{mnpl} = \langle mn| \frac{e^2}{|\mathbf{x} - \mathbf{x}'|} |pl\rangle = e^2 \int d\mathbf{x} \int d\mathbf{x}' \frac{\phi_m^*(\mathbf{x})\phi_n^*(\mathbf{x}')\phi_l(\mathbf{x}')\phi_p(\mathbf{x})}{|\mathbf{x} - \mathbf{x}'|} \delta_{\sigma_m \sigma_p} \delta_{\sigma_n \sigma_l}$$

$$(1.34)$$

You can check how these expressions are obtained by introducing the definition for the field operators in Eqs. (1.28) and (1.29). The Feynman diagram for the Coulomb matrix element, V_{mnpl}, is:

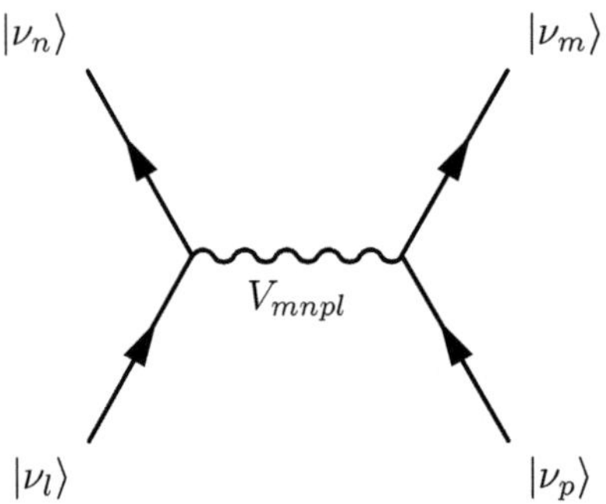

Exercise I.1: Hamiltonian of an interacting electron gas.
The hamiltonian of an interacting electron gas can be expressed in second quantization form as:

$$H = \sum_\sigma \int d\mathbf{x} \Psi_\sigma^\dagger(\mathbf{x}) \left(-\frac{\hbar^2 \nabla^2}{2m}\right) \Psi_\sigma(\mathbf{x}) + \frac{1}{2} \int d\mathbf{x} \int d\mathbf{x}' \frac{e^2}{|\mathbf{x} - \mathbf{x}'|} : n(\mathbf{x})n(\mathbf{x}') :,$$

where, $n(\mathbf{x}) = \sum_\sigma \Psi_\sigma^\dagger(\mathbf{x})\Psi_\sigma(\mathbf{x})$, is the particle density at $\mathbf{x}$ and $\Psi_\sigma^\dagger(\mathbf{x})$ is the field operator creating a fermion with spin σ at position $\mathbf{x}$. The symbol $::$ denotes normal order of the field operators. Using the complete, normalized basis of momentum eigenstates $|\mathbf{p}, \sigma\rangle$ defining the one-particle Hilbert space, calculate the matrix elements in the occupation number basis: $|n_{\mathbf{p}_1\sigma_1}, n_{\mathbf{p}_2\sigma_2}, \ldots\rangle$ of the associated multi-particle Fock space.

(i) Show that the matrix elements conserve the total momentum of the system, i.e., the sum of outgoing momenta is equal to the sum of ingoing momenta.

(ii) Provide the final expression of H in second quantization.

(iii) Explain why normal ordering is needed to correctly describe the Coulomb interaction.

1.4 Hubbard Model

We have already introduced a simplified model for the electron interacting gas. We now consider a model of interacting electrons and localized magnetic moments on a lattice. The Hubbard model is the standard model in strongly correlated systems relevant to high-T_c superconductivity, organic superconductivity and magnetism in transition metal.

We consider a lattice of atoms in which electrons are almost localized due to the small spatial extension of the atomic wavefunctions with respect to the lattice parameter. Hence, we should use localized atomic orbitals as an appropriate basis set to describe the band structure of this system. Take a single $1s$-orbital at each site i, then $\{\Psi_{1s,i}(\mathbf{x}) = \phi_{1s}(\mathbf{x} - \mathbf{x}_i)$. In this basis the field operator reads:

$$\Psi_\sigma^\dagger = \sum_j \phi_{1s}^*(\mathbf{x} - \mathbf{x}_j)c_{j\sigma}^\dagger \tag{1.35}$$

Using these field operators in the hamiltonian

$$H = \sum_\alpha \int d\mathbf{x}\Psi_\alpha^\dagger(\mathbf{x})h_{oe}(\mathbf{x})\Psi_\alpha(\mathbf{x}) + \frac{1}{2}\sum_{\alpha,\beta} \int d\mathbf{x}d\mathbf{x}' \Psi_\alpha^\dagger(\mathbf{x})\Psi_\beta^\dagger(\mathbf{x}')v_{ee}(\mathbf{x} - \mathbf{x}')\Psi_\beta(\mathbf{x}')\Psi_\alpha(\mathbf{x}). \tag{1.36}$$

with:

$$h_{oe}(\mathbf{x}) = -\frac{\hbar^2\nabla^2}{2m} + V(\mathbf{x}),$$

$$v_{ee}(\mathbf{x} - \mathbf{x}') = \frac{e^2}{|\mathbf{x} - \mathbf{x}'|}. \tag{1.37}$$

This would lead to the hamiltonian in second quantized form:

$$H = \sum_\alpha \int d\mathbf{x}T_{ij}c_{i\alpha}c_{j\alpha} + \frac{1}{2}\sum_{i,j,k,l;\alpha,\beta} \langle ij|v_{ee}|lk\rangle c_{i\alpha}^\dagger c_{j\beta}^\dagger c_{k\beta}c_{l\alpha} \tag{1.38}$$

where the matrix elements read:

$$T_{ij} = \langle i|h_{oe}|j\rangle = \int d\mathbf{x}\phi_i^*(\mathbf{x})h_{oe}(\mathbf{x})\phi_j(\mathbf{x})V_{ij,lk}\langle ij|v_{ee}|lk\rangle$$

$$= \int d\mathbf{x}d\mathbf{x}'\phi_i^*(\mathbf{x})\phi_l(\mathbf{x})v_{ee}(\mathbf{x} - \mathbf{x}')\phi_j^*(\mathbf{x}')\phi_k(\mathbf{x}'). \tag{1.39}$$

In the simplest version of the Hubbard model one retains only the hopping matrix elements between nearest neighbor sites and onsite Coulomb repulsion between two electrons on a given site. In this case, the model parameters reduce to:

$$T_{ij} = \begin{cases} \epsilon, & \text{if } i = j \\ -t, & \text{if } i, j \text{ n. n.} \\ 0, & \text{otherwise} \end{cases} \tag{1.40}$$

and:

$$V_{ij,lk} = \begin{cases} U, & \text{if } i = j = l = k \\ 0, & \text{otherwise} \end{cases} \tag{1.41}$$

Under these simplifications we recover the Hubbard model:

$$H = \epsilon \sum_{j\sigma} c_{j\sigma}^{\dagger} c_{j\sigma} - t \sum_{\langle ij,\sigma \rangle} c_{i\sigma} c_{j\sigma} + U \sum_{j} n_{j\uparrow} n_{j\downarrow}, \tag{1.42}$$

since:

$$\frac{U}{2} \sum_{j,\sigma,\sigma'} c_{j\sigma}^{\dagger} c_{j\sigma'}^{\dagger} c_{j\sigma'} c_{j\sigma} = \frac{U}{2} \sum_{j} (c_{j\uparrow}^{\dagger} c_{j\downarrow}^{\dagger} c_{j\downarrow} c_{j\uparrow} + c_{j\downarrow}^{\dagger} c_{j\uparrow}^{\dagger} c_{j\uparrow} c_{j\downarrow}) \tag{1.43}$$

since $c_{j\sigma}^2 = 0$ the Coulomb interaction between two electrons with the same spin, σ, vanishes due to the Pauli exclusion principle (two fermions cannot be in the same state with exactly the same quantum numbers). This apparently simple model describes a very important physical phenomenon at half-filling: the Mott transition. While for $U \to 0$ the system is a metal consisting on a half-filled band, in the limit $U \gg t$ electrons localize forming a Mott insulator. While the Hubbard model in 1D has been solved exactly, there is no exact solution yet in 2D or 3D.

Exercise I.2: Hubbard dimer
Consider a half-filled Hubbard dimer:

$$H = -t \sum_{\sigma} (c_{1\sigma}^{+} c_{2\sigma} + c.c.) + U \sum_{i=1}^{2} n_{i\uparrow} n_{i\downarrow}$$

(i) Obtain all eigenenergies and eigenstates of the system in the $S_{tot}^z = 0$ sector. Plot the energies as a function of U/t.

(ii) Show that the ground state of the system is a singlet which reads:

$$|\Psi_0 >= A(|\uparrow, \downarrow> - |\downarrow, \uparrow> + \beta(|0, \uparrow\downarrow> + |\uparrow\downarrow, 0 >)),$$

show that in the limit $U \gg t$, $\beta \to 0$, the ground state reduces to:

$$|\Psi_0 >\approx \frac{1}{\sqrt{2}}(|\uparrow, \downarrow> - |\downarrow, \uparrow>).$$

(iii) In the $U \gg t$ limit show that to $O(t^2/U)$ the model maps onto a Heisenberg model:

$$H = J\mathbf{S}_1 \cdot \mathbf{S}_2$$

with: $J = 4t^2/U > 0$. Find the spectra of this model and compare with the one found in (i) when $U \gg t$.

1.5 Heisenberg Model

In the large U/t limit of the Hubbard model, virtual excursions of electrons to neighbor sites lead to an AF exchange. Hence, the spin degrees of freedom in the Mott localized state are governed by an AF Heisenberg model:

$$H_{Hubbard} \rightarrow H_{Heisenberg} = J \sum_{\langle ij \rangle} \mathbf{S}_i \cdot \mathbf{S}_j, \tag{1.44}$$

with $J = \frac{4t^2}{U} > 0$. (see Exercise I.2).

Spin operators satisfy the algebra:

$$[S_i^{\alpha}, S_j^{\beta}] = i \epsilon^{\alpha\beta\gamma} S_i^{\gamma} \delta_{ij}. \tag{1.45}$$

which is different from standard boson commutation relations since S_i^{γ} is not a c-number. The Heisenberg model can be analyzed just based on these commutation relations. However, spin-1/2 operators can be represented, with no loss of generality, by the Pauli matrices:

$$S^x = \frac{1}{2} \begin{pmatrix} 0 & 1 \\ 1 & 0 \end{pmatrix}, S^y = \frac{1}{2} \begin{pmatrix} 0 & -i \\ i & 0 \end{pmatrix}, S^z = \frac{1}{2} \begin{pmatrix} 1 & 0 \\ 0 & -1 \end{pmatrix}. \tag{1.46}$$

In analogy with orbital momentum operators to introduce lowering and raising operators:

$$\begin{aligned} S_i^+ &= S_i^x + i S_i^y \\ S_i^- &= S_i^x - i S_i^y \end{aligned} \tag{1.47}$$

which increase (lower) the total z-component of the spin. These operators satisfy the commutation relations:

$$[S_i^+, S_j^-] = 2 S_j^z \delta_{ij}, [S_i^z, S_j^+] = S_j^+ \delta_{ij}, [S_i^z, S_j^-] = -S_j^- \delta_{ij}. \tag{1.48}$$

A generalized version of the Heisenberg model allows describing exchange anisotropies:

$$H = \sum_{\langle ij \rangle} (J_x S_i^x S_j^x + J_y S_i^y S_j^y + J_z S_i^z S_j^z). \tag{1.49}$$

with J_x, J_y, J_z arbitrary. Different important models are:

(a) Anisotropic Heisenberg: If $J_x = J_y = J_\perp \neq 0$ and $J_z \neq 0$.
(b) Ising model: If $J_x = J_y = 0 \neq 0$ and $J_z \neq 0$.
(c) XY model: If $J_x = J_y = J_\perp \neq 0$ and $J_z = 0$.

Note that since two spin operators anticommute on the same site:

$$\{S_i^\alpha, S_i^\beta\} = \frac{\delta^{\alpha\beta}}{2} \tag{1.50}$$

the ladder operators satisfy:

$$\{S_i^+, S_i^-\} = 1. \tag{1.51}$$

So spins on the same sites behave as fermions. However, on different sites the spin operators commute. So spin operators are NOT exactly like fermion creation and destruction operators. Due to the complex commutation relations of spin operators, even a quadratic spin hamiltonian such as the Heisenberg model is non-trivial to solve. This is in contrast to quadratic fermion hamiltonians which can be diagonalized straightforwardly. Hence, spin models can lead to very interesting states which are not well understood yet. One exotic phase is the quantum spin liquid, a magnetically disordered state of antiferromagnetically coupled spins on a lattice which occurs at zero temperature!

Jordan-Wigner transformation in 1D
The approximate analogy between spins and fermions was used by Jordan and Wigner to introduce an exact mapping of spin operators onto spinless fermions in 1D. If we have a chain of $S = 1/2$ spins located at $i = 1, 2, 3, \ldots$ numbered from left to right, the spin operator at site i is given by:

$$S_i^z = f_i^\dagger f_i - \frac{1}{2}, \; S_i^+ = f_i^+ e^{i\pi \sum_{l<i} n_l}, \; S_i^- = e^{-i\pi \sum_{l<i} n_l} f_i,$$

where n_l is the fermion occupation number of site l, $n_l = 0, 1$. The phases count the number of fermions to the left of the spin at site i. This transformation satisfies all the commutation and anticommutation relations of the spin operators.

Exercise I.3: Jordan-Wigner transformation

(i) Use the Jordan-Wigner transformation to show that the one dimensional anisotropic XY model:

$$H = -\sum_i (J_x S_i^x S_{i+1}^x + J_y S_i^y S_{i+1}^y)$$

can be written as:

$$H = \sum_i [-t(f_{i+1}^\dagger f_i + H.c.) + \Delta(f_{i+1}^\dagger f_i^\dagger + H.c.)],$$

where: $t = (J_x + J_y)/4$ and $\Delta = (J_y - J_x)/4$.

(ii) Calculate the excitation spectrum for this model and sketch your results. Comment specifically on the two cases $J_x = J_y$ and $J_y = 0$.

This exercise is set and solved in [6].

Using this transformation the 1D XY model can be solved exactly. This model can be expressed as:

$$H^{XY} = -J_\perp \sum_i (S_i^+ S_{i+1}^- + S_i^- S_{i+1}^+). \tag{1.52}$$

Using the Jordan-Wigner transformation it can be expressed as free spinless fermions in a chain:

$$H^{XY} = -J_\perp \sum_i (f_i^+ f_{i+1} + f_{i+1}^+ f_i). \tag{1.53}$$

which can be diagonalized exactly by transforming to momentum space. The diagonalized hamiltonian reads:

$$H^{XY} = -2J_\perp \sum_k \gamma_k f_k^+ f_k. \tag{1.54}$$

where $\gamma_k = \cos(k)$. Hence, the excitation spectrum of the XY model in 1D is equivalent to that of a system of free spinless fermions! The ground state of the model is spin disordered which is interpreted as a quantum spin liquid. This is unlike classical antiferromagnets whose ground state is AF ordered and the elementary excitations are spin waves. At present, there is no known equivalent Wigner transformation for 2D or 3D spin systems.

1.6 Anderson Model of Magnetic Impurities in Metals

The above Hubbard and Heisenberg refer to homogeneous interacting electron and spin systems on a lattice. Impurity models refer to systems in which an atom or magnetic moment is immersed in an homogeneous electron gas. The Hilbert space consists of plane waves, $\{\phi_\mathbf{k}(\mathbf{x})\}$ describing the metallic host and a localized atomic orbital, $\phi_d(\mathbf{x}) = \phi_d(\mathbf{x} - \mathbf{R}_n)$, at $\mathbf{R}_n$ describing a single localized orbital in the atom, $\{\phi_\mathbf{k}(\mathbf{x}), \phi_d(\mathbf{x})\}$. The field operators in this basis read:

$$\Psi_\sigma^+(\mathbf{x}) = \sum_\mathbf{k} \phi_\mathbf{k}^*(\mathbf{x}) c_{\mathbf{k}\sigma}^+ + \phi_d^*(\mathbf{x}) c_{d\sigma}^+ \tag{1.55}$$

And the hamiltonian in second quantization can be obtained from:

$$H = \sum_\sigma \int d\mathbf{x} \Psi_\sigma^\dagger(\mathbf{x}) h_{oe}(\mathbf{x}) \Psi_\sigma(\mathbf{x}) + \frac{1}{2} \sum_{\alpha\beta} \int d\mathbf{x} \int d\mathbf{x}' \Psi_\alpha^\dagger(\mathbf{x}) \Psi_\beta^\dagger(\mathbf{x}') v_{ee}(\mathbf{x}) \Psi_\beta(\mathbf{x}') \Psi_\alpha(\mathbf{x}) \tag{1.56}$$

by substituting the field operators giving:

$$H = \sum_{\mathbf{k}\sigma} \epsilon_{\mathbf{k}} c_{\mathbf{k}\sigma}^{\dagger} c_{\mathbf{k}\sigma} + \sum_{\sigma} \epsilon_d c_{d\sigma}^{\dagger} c_{d\sigma} + \sum_{\mathbf{k},\mathbf{k}',\sigma} V_{\mathbf{k},\mathbf{k}'} c_{\mathbf{k}\sigma}^{\dagger} c_{\mathbf{k}'\sigma}$$
$$+ \sum_{bfk\sigma} V_{\mathbf{k}d}(c_{\mathbf{k}\sigma}^{\dagger} c_{d\sigma} + c_{d\sigma}^{\dagger} c_{\mathbf{k}\sigma}) + U n_{d\uparrow} n_{d\downarrow}. \tag{1.57}$$

where:

$$\epsilon_{\mathbf{k}} = \int d\mathbf{x} \phi_{\mathbf{k}}^{*}(\mathbf{x})(-\frac{\hbar^2}{2m}\nabla^2 + V(\mathbf{x}))\phi_{\mathbf{k}}(\mathbf{x}),$$
$$V_{\mathbf{k},\mathbf{k}'} = \int d\mathbf{x} \phi_{\mathbf{k}}^{*}(\mathbf{x})(-\frac{\hbar^2}{2m}\nabla^2 + V(\mathbf{x}))\phi_{\mathbf{k}}'(\mathbf{x}),$$
$$U = \int d\mathbf{x} \int d\mathbf{x}' |\phi_d(\mathbf{x} - \mathbf{R}_n)|^2 \frac{e^2}{|\mathbf{x} - \mathbf{x}'|} |\phi_d(\mathbf{x}' - \mathbf{R}_n)|^2. \tag{1.58}$$

So the Anderson model keeps only the onsite Coulomb repulsion on the impurity, U. Although other terms are present, U is the largest contribution since the d-orbital is the most spatially localized orbital in the model. The Anderson model is a very good model to describe the physics of atom impurities in metals. The $V_{\mathbf{k}\mathbf{k}'}$ can be absorbed in the plane wave basis sets by solving the scattering problem so ti can be dropped with no change in the relevant physics. Hence, we reach the Anderson model:

$$H_{AM} = \sum_{\mathbf{k}\sigma} \epsilon_{\mathbf{k}} c_{\mathbf{k}\sigma}^{\dagger} c_{\mathbf{k}\sigma} + \sum_{\sigma} \epsilon_d c_{d\sigma}^{\dagger} c_{d\sigma} + \sum_{\mathbf{k}\sigma} V_{\mathbf{k}\sigma}(c_{\mathbf{k}\sigma}^{\dagger} c_{d\sigma} + c.c.) + U n_{d\uparrow} n_{d\downarrow}.$$
$$\tag{1.59}$$

In the limit $U >> t$, the Anderson model reduces to the Kondo model which reads:

$$H_{KM} = \sum_{\mathbf{k}\sigma} \epsilon_{\mathbf{k}} c_{\mathbf{k}\sigma}^{\dagger} c_{\mathbf{k}\sigma} + \sum_{\sigma} \epsilon_d n_{d\sigma}$$
$$+ \sum_{\mathbf{k},\mathbf{k}'} J_{\mathbf{k}\mathbf{k}'}(S_d^{+} c_{\mathbf{k}\downarrow} c_{\mathbf{k}'\uparrow} + S_d^{-} c_{\mathbf{k}\uparrow}^{\dagger} c_{\mathbf{k}'\downarrow} + S_d^{z}(c_{\mathbf{k}\uparrow} c_{\mathbf{k}'\uparrow} - c_{\mathbf{k}\downarrow} c_{\mathbf{k}'\downarrow})), \tag{1.60}$$

which is the starting point to study magnetic impurities in metals. Kondo realized that spin-flip processes could explain the resistance minimum observed in metals with dilute magnetic impurities.

Exercise I.4:   Kitaev spin model

The Kitaev spin model (Kitaev 2006) describes the interaction of $S = 1/2$ on a honeycomb lattice. can be solved exactly by expressing spin operators in terms of Majorana fermions. As shown by Kitaev the exact solution, which we analyze in this exercise, is a quantum spin liquid. The model reads:

$$H = -J_x \sum_{x-link} S_i^x S_j^x - J_y \sum_{y-link} S_i^y S_j^y - J_z \sum_{z-link} S_i^z S_j^z, \qquad (1.61)$$

where the sums are over the x, y and z-links of the honeycomb lattice shown below.

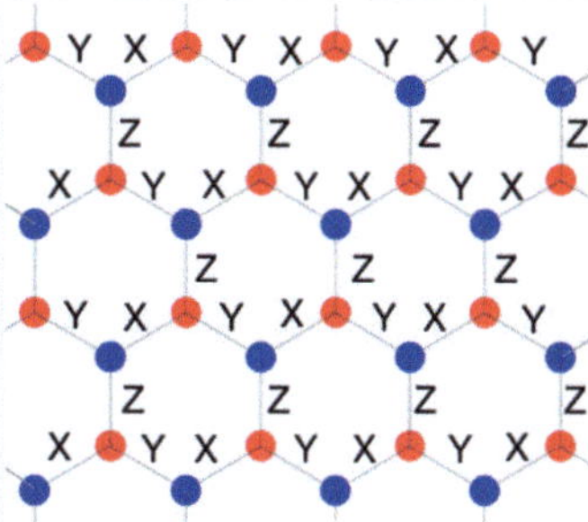

Spin operators can be represented through Schwinger fermions:

$$\mathbf{S}_i = \frac{1}{2} \sum_{\alpha\beta} f_{i\sigma}^\dagger \boldsymbol{\sigma}_{\alpha\beta} f_{i\beta}. \qquad (1.62)$$

under the constraints:

$$\sum_\sigma f_{i\sigma}^\dagger f_{i\sigma} = 1, \; f_{i\uparrow}^\dagger f_{i\downarrow}^\dagger = f_{i\uparrow} f_{i\downarrow} = 0, \qquad (1.63)$$

so that the original Hilbert space of the spins is recovered.

(i) Show that by expressing Schwinger fermions in terms of four Majorana
    fermions at each site:

$$f_{j\uparrow} = \frac{1}{2}(b_j^x - ib_j^y)$$

$$f_{j\downarrow} = \frac{1}{2}(-b_j^z + ic_j). \tag{1.64}$$

the spin operators reduce to:

$$S_j^\alpha = \frac{i}{2}b_j^\alpha c_j \tag{1.65}$$

making use of the constraints. Obtain an expression of the original Kitaev
spin model in terms of the Majorana operators.

(ii) Show that the model obtained becomes quadratic and so it can be diag-
     onalized exactly. Obtain the ground state of the model for $K = J_x = J_y = J_z$. Discuss the solution.

Part I
Equilibrium Many-Body Techniques

Chapter 2
Introduction to Green-Function Methods

It is not possible, in general, to find the wavefunction, $\Psi(\mathbf{x}_1, \mathbf{x}_2, ...\mathbf{x}_N)$ of a system of $N \approx 10^{23}$ particles by solving straightforwardly the Schrödinger equation:

$$\left(-\frac{\hbar^2}{2m} \sum_j \nabla_j^2 + \sum_{i<j} V_{e-e}(\mathbf{x}_i - \mathbf{x}_j) + \sum_j V(\mathbf{x}_j)) \right)\Psi = i\hbar \frac{\partial \Psi}{\partial t}. \qquad (2.1)$$

Also, experiments measure elementary excitations of these systems. For instance, in photoemission experiments an incident photon causes an electron to be ejected from the system. To a good approximation, the intensity of the ejected electrons at $\mathbf{k}, \omega$ is given by:

$$I(\mathbf{k}, \omega) \propto -Im G^R(\mathbf{k}, \omega), \qquad (2.2)$$

where $G^R(\mathbf{k}, \omega)$ is the Fourier transform of a retarded Green function:

$$G^R(\mathbf{k}, t, t') = G^R(\mathbf{k}, t - t') = -i\theta(t - t')\langle\{\Psi_\mathbf{k}(t), \Psi_\mathbf{k}(t')\}\rangle, \qquad (2.3)$$

where $\langle\rangle$ denotes the thermal average over all spectra of quantum many body states.

In general, as we will see, we can also obtain the response of a system to a small external potential from the retarded response functions of the form:

$$\chi_{ij} = -i\theta(t - t')\langle[A_i(t), A_j(t')]\rangle. \qquad (2.4)$$

Our final aim will be to evaluate these Green functions directly connected to experiments. The basic building block to construct Green functions is the field operator introduved earlier.

J. Merino and A. L. Yeyati, *Many-Body Techniques in Condensed Matter Physics*,
UNITEXT for Physics, https://doi.org/10.1007/978-3-031-55143-7_2

2.1   Time Evolution

We would like to find the evolution of the system with time to be able to evaluate the Green functions. The main goal here will be to find the time-evolution of the field operators themselves. We start by revising the different ways to time evolve the system in one-particle quantum mechanics which are formally equivalent:

Schrödinger picture
Time-evolution is encoded in the wavefunction only, $|\Psi_s(t)\rangle$:

$$\frac{i\partial}{\partial t}|\Psi_s(t)\rangle = H|\Psi_s(t)\rangle, \tag{2.5}$$

which has the formal solution:

$$|\Psi_s(t)\rangle = e^{-iH(t-t_0)}|\Psi_s(t_0)\rangle. \tag{2.6}$$

The time evolution operator is defined as:

$$U(t, t_0) = e^{-iH(t-t_0)} \tag{2.7}$$

which time-evolves the wavefunction from $|\Psi_s(t_0)\rangle$ to $|\Psi_s(t)\rangle$.

The average value of an observable at time t reads:

$$\langle O_s(t)\rangle = \langle\Psi_s(t)|O_s|\Psi_s(t)\rangle = \langle\Psi_s(t_0)|e^{iH(t-t_0)}O_s e^{-iH(t-t_0)}|\Psi_s(t_0)\rangle. \tag{2.8}$$

Heisenberg picture
We make a unitary transformation to new state vectors $|\Psi_H\rangle$ and observables O_H:

$$\begin{aligned}
|\Psi_H\rangle &\equiv e^{iH(t-t_0)}|\Psi_s(t)\rangle = |\Psi_s(t_0)\rangle \\
O_H(t) &= e^{iH(t-t_0)}O_s e^{-iH(t-t_0)}
\end{aligned} \tag{2.9}$$

So only equations depend on time. The observables are just equivalent to Schrödingers picture:

$$\langle\Psi_H|O_H(t)|\Psi_H\rangle = \langle\Psi_s(t_0)|e^{iH(t-t_0)}O_s e^{-iH(t-t_0)}|\Psi_s(t_0)\rangle. \tag{2.10}$$

Instead of Schrödingers equation of motion we have Heisenbergs equation of motion:

$$i\frac{d}{dt}O_H(t) = i^2 H e^{iH(t-t_0)}O_s e^{-iH(t-t_0)} + e^{iH(t-t_0)}O_s(-i^2 H)e^{-iH(t-t_0)}, \tag{2.11}$$

or equivalently:

$$i\frac{d}{dt}O_H = [O_H, H], \tag{2.12}$$

with $O_H(t_0) = O_s$.

2.2 Time-Ordered or Causal Green's Functions at $T = 0$

In QFT, the one-particle Green function is the fundamental quantity characterizing the microscopic properties of a system. It is defined as:

$$G_{\alpha\beta}(x, x') \equiv -i \langle T[\Psi_{H\alpha}(x)\Psi_{H\beta}^{\dagger}(x')]\rangle, \tag{2.13}$$

where $x = (\mathbf{x}, t)$ or $x' = (\mathbf{x}', t')$ and α, β are spin indices. For the time being we are assuming that $T = 0$ so the average $\langle\rangle = \langle\Psi_G|\Psi_G\rangle$. These Green functions introduced by Feynman in the context of particle physics are called time-ordered or causal Green functions and are amenable to a perturbation expansion. They are directly relevant for analyzing ARPES experiments!

The definition of time ordering is:

$$T[\Psi_{\alpha}^{H}(\mathbf{x}, t)\Psi_{\beta}^{H}(\mathbf{x}', t'))] = \Psi_{H\alpha}(\mathbf{x}, t)\Psi_{H\beta}^{\dagger}(\mathbf{x}', t')\theta(t - t')$$
$$\pm \Psi_{H\beta}^{\dagger}(\mathbf{x}', t')\Psi_{H\alpha}(\mathbf{x}, t)\theta(t' - t),$$

where $+$ refers to bosons and $-$ to fermions. The Green's function gives the probability amplitude that an added particle at $\mathbf{x}'$ at time t' is found at position $\mathbf{x}$ at a later time t or that an added hole at ($\mathbf{x}$ at time t is found at position $\mathbf{x}'$ at time t'. Hence, it describes the propagation of an electron or a hole through the system in the presence of interactions. From now we will drop the subscript H in the Heisenberg fields: $\Psi^{H}(x) \rightarrow \Psi(x)$ and $G_{\alpha\beta} = G\delta_{\alpha\beta}$ is diagonal in the spin indices. We will consider fermions and bosons. For bosons we will study phonons which do not present Bose condensation.

We will mostly deal with homogeneous in time and spatially infinite systems which depend only on spatial coordinate differences, $\mathbf{x} - \mathbf{x}'$ and time differences, $t - t'$. Hence, we can describe/represent the Green function as a Fourier integral:

$$G(x, x') = G(x - x') = G(\mathbf{x} - \mathbf{x}', t - t') = \int \frac{d^4k}{(2\pi)^4}e^{ik(x-x')}G(k)$$
$$= \int \frac{d\mathbf{k}}{(2\pi)^3} \int \frac{d\omega}{(2\pi)}e^{i[\mathbf{k}(\mathbf{x}-\mathbf{x}')-\omega(t-t')]}G(\mathbf{k}, \omega). \tag{2.14}$$

Finally, the momentum dependent Green function reads:

$$G(\mathbf{k}, t - t') = \int \frac{d\omega}{2\pi}G(\mathbf{k}, \omega)e^{-i\omega(t-t')}, \tag{2.15}$$

which we will use at different points of the course.

2.3 Causal Green Function for Free Fermions

As an example let's consider the Green function in a free fermion gas:

$$H = H^0 - \mu N = \sum_{\mathbf{k},\sigma} \epsilon_{\mathbf{k}} c^{\dagger}_{\mathbf{k}\sigma} c_{\mathbf{k}\sigma}, \tag{2.16}$$

where $\epsilon_{\mathbf{k}} = \epsilon^0_{\mathbf{k}} - \mu$. So the hamiltonian is expressed in the grand canonical ensemble. Solving the Heisenberg equation of motion for $c_{\mathbf{k}\sigma}(t)$ we have:

$$i\frac{\partial}{\partial t} c_{\mathbf{k}\sigma}(t) = [c_{\mathbf{k}\sigma}(t), H] = e^{iHt} c_{\mathbf{k}\sigma} e^{-iHt} H - H e^{iHt} c_{\mathbf{k}\sigma} e^{-iHt} =$$
$$= e^{iHt} [c_{\mathbf{k}\sigma}, H] e^{-iHt} = \epsilon_{\mathbf{k}} c_{\mathbf{k}\sigma}(t), \tag{2.17}$$

since $[c_{\mathbf{k}\sigma}, H] = \epsilon_{\mathbf{k}} c_{\mathbf{k}\sigma}$ with solution:

$$c_{\mathbf{k}\sigma}(t) = e^{-i\epsilon_{\mathbf{k}}t} c_{\mathbf{k}\sigma},$$
$$c^{\dagger}_{\mathbf{k}\sigma}(t) = e^{i\epsilon_{\mathbf{k}}t} c^{\dagger}_{\mathbf{k}\sigma}. \tag{2.18}$$

where we have included the solution of $c^{\dagger}_{\mathbf{k}\sigma}(t)$ for completeness.

We now introduce this time-dependent operators in the definition of the Green function. For $t > t'$ (forward propagation) we have:

$$G_{\sigma\sigma'}(\mathbf{k}, t - t') = -i\langle \Psi_G | c_{\mathbf{k}\sigma}(t) c^{\dagger}_{\mathbf{k}\sigma'}(t') | \Psi_G \rangle = -i e^{-i\epsilon_{\mathbf{k}}(t-t')} \langle \Psi_G | c_{\mathbf{k}\sigma} c^{\dagger}_{\mathbf{k}\sigma'} | \Psi_G \rangle$$
$$= -i e^{-i\epsilon_{\mathbf{k}}(t-t')} \delta_{\sigma\sigma'} (1 - \langle n_{\mathbf{k}\sigma} \rangle), \tag{2.19}$$

where we have used: $|\Psi_G\rangle = \prod_{|\mathbf{k}'|<k_F} c^{\dagger}_{\mathbf{k}'\sigma}|0\rangle$, the ground state of the Fermi gas. Note that $\langle n_{\mathbf{k}\sigma} \rangle = \langle \Psi_G | n_{\mathbf{k}\sigma} | \Psi_G \rangle = \theta(k_F - |\mathbf{k}|)$ Similarly, for $t < t'$ (backward propagation) we find:

$$G_{\sigma\sigma'}(\mathbf{k}, t - t') = -i\langle \Psi_G | c_{\mathbf{k}\sigma}(t) c^{\dagger}_{\mathbf{k}\sigma'}(t') | \Psi_G \rangle = i e^{-i\epsilon_{\mathbf{k}}(t-t')} \delta_{\sigma\sigma'} \langle n_{\mathbf{k}\sigma} \rangle. \tag{2.20}$$

In summary:

$$G_{\sigma}(\mathbf{k}, t) = -i(\theta(k - k_F)\theta(t) e^{-i\epsilon_{\mathbf{k}}t} - \theta(k_F - k)\theta(-t) e^{i\epsilon_{\mathbf{k}}t}) \tag{2.21}$$

The frequency decomposition can be obtained by Fourier transform:

$$G(\mathbf{k}, \omega) = -i\{\theta(k - k_F) \int_{-\infty}^{\infty} \theta(t)e^{i(\omega - \epsilon_\mathbf{k})t}\,dt - \theta(k_F - k) \int_{-\infty}^{\infty} \theta(-t)e^{-i(\omega - \epsilon_\mathbf{k})t}\,dt\}.$$

$$(2.22)$$

We can perform time integration:

$$\lim_{\eta \to 0^+} \int_{0}^{\infty} e^{i(\omega - \epsilon_\mathbf{k} + i\eta)t}\,dt = \frac{1}{i(i\eta + (\omega - \epsilon_\mathbf{k}))} e^{-\eta t} e^{i(\omega - \epsilon_\mathbf{k})t} \big|_{0}^{\infty} = i \lim_{\eta \to 0^+} \frac{1}{\omega - \epsilon_\mathbf{k} + i\eta}.$$

$$(2.23)$$

which gives the final expression for $G(\mathbf{k}, \omega)$:

$$G(\mathbf{k}, \omega) = \frac{\theta(k - k_F)}{\omega - \epsilon_\mathbf{k} + i\eta} + \frac{\theta(k_F - k)}{\omega - \epsilon_\mathbf{k} - i\eta} = \frac{1}{\omega - \epsilon_\mathbf{k} + i\eta \, sign(k - k_F)} \qquad (2.24)$$

which has simple poles in the ω-complex plane at $\omega = \epsilon_\mathbf{k} + i\eta$ for $k < k_F$ (holes) and $\omega = \epsilon_\mathbf{k} - i\eta$ for $k > k_F$ (electrons) as shown in Fig. 2.1

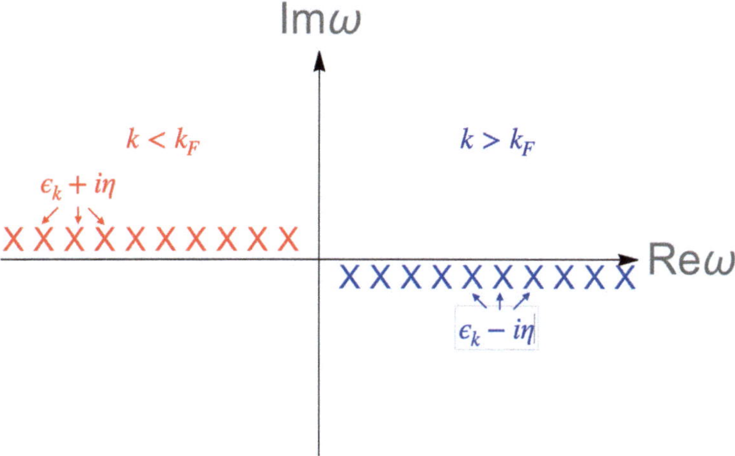

Fig. 2.1 Pole structure of the causal Green function of a free electron in the complex frequency plane. Single excitation energies are referred to the Fermi energy

> **Exercise I.5: Green function of phonons**
> Obtain the propagator of a gas of non-interacting phonons described by:
>
> $$H = \sum_{\mathbf{q}} \omega_{\mathbf{q}} \left[b_{\mathbf{q}}^{\dagger} b_{\mathbf{q}} + \frac{1}{2} \right],$$
>
> where the physical displacement field operator, $\phi(\mathbf{x})$ is related to a sum of creation and destruction operators:
>
> $$\phi(\mathbf{x}) = \int \frac{d^3 q}{(2\pi)^3} \phi(\mathbf{q}) e^{i\mathbf{q}\cdot\mathbf{x}}$$
>
> where: $\phi(\mathbf{q}) = \sqrt{\frac{\hbar}{2m\omega_{\mathbf{q}}}}(b_{\mathbf{q}} + b_{-\mathbf{q}}^{\dagger})$. The phonon propagator in momentum space is defined as: $D(\mathbf{q}, t) = -i\langle T[\phi(\mathbf{q}, t)\phi(-\mathbf{q}, 0)]\rangle$, where the ground state of the system is the vacuum of bosons. Obtain the frequency dependence: $D(\mathbf{q}, \nu)$ by performing a Fourier transform of $D(\mathbf{q}, t)$. Discuss the physical meaning of the poles of $D(\mathbf{q}, \nu)$.

2.4 Physical Observables

We can obtain static and dynamical quantities from the Green function such as:

1. Average values of one-electron operators.
2. Ground state energy.
3. One-electron excitations of the system (as we have seen for the electron gas).

Regarding 1. we have, for example that the average number of particles at a given point, $\mathbf{x}$, reads:

$$\langle n(\mathbf{x})\rangle = \sum_{\sigma}\langle \Psi_{\sigma}^{\dagger}(\mathbf{x})\Psi_{\sigma}(\mathbf{x})\rangle = -\sum_{\sigma}\langle \Psi_{G}|T[\Psi_{\sigma}(\mathbf{x}, 0)\Psi_{\sigma}(\mathbf{x}', 0^{+})]|\Psi_{G}\rangle|_{\mathbf{x}'\to\mathbf{x}} =$$

$$-i(2S+1)G(\mathbf{x}, 0^{-})|_{\mathbf{x}=0} = -i(2S+1)lim_{\mathbf{x}'\to\mathbf{x}, t'\to t+0^{+}} G(\mathbf{x} - \mathbf{x}', t - t')$$

$$\tag{2.25}$$

The kinetic energy is:

$$\langle T(\mathbf{x})\rangle = \sum_{\sigma}\langle \Psi_{\sigma}^{\dagger}(\mathbf{x})(-\frac{\hbar^2}{2m}\nabla_x^2\Psi_{\sigma}(\mathbf{x}))\rangle = \frac{\hbar^2}{2m}\nabla_x^2\sum_{\sigma}\langle T[\Psi_{\sigma}^{\dagger}(\mathbf{x})\Psi_{\sigma}(\mathbf{x})]\rangle =$$

$$i(2S+1)\frac{\hbar^2\nabla_x^2}{2m}G(\mathbf{x} - \mathbf{x}', t - t')|_{\mathbf{x}'\to\mathbf{x}, t'\to t+0^{+}} = i(2S+1)\frac{\hbar^2\nabla_x^2}{2m}G(\mathbf{x}, 0^{-})|_{\mathbf{x}=0}$$

$$\tag{2.26}$$

Exercise I.6: Green functions and ground state properties of a non-interacting electron gas

In the same way as we defined the momentum dependent causal Green function we can define a retarded Green function:

$$G_\sigma^R(\mathbf{k}, t - t') = -i\theta(t - t')\langle\{c_{\mathbf{k}\sigma}(t), c_{\mathbf{k}\sigma}^\dagger(t')\}\rangle.$$

Here, we consider a noninteracting electron gas:

$$H_0 - \mu N = \sum_{\mathbf{k}\sigma} \epsilon_{\mathbf{k}} c_{\mathbf{k}\sigma}^\dagger c_{\mathbf{k}\sigma},$$

where: $\epsilon_{\mathbf{k}} = E_{\mathbf{k}} - \mu$ is the energy of non-interacting electrons referred to the Fermi energy $\mu(T = 0) = E_F$.

(i) Compute the explicit time dependence of the retarded Green function.

(ii) Fourier transforming the results in (a) show that:

$$G_\sigma^{(0)R}(\mathbf{k}, \omega) = \frac{1}{\omega - \epsilon_{\mathbf{k}} + i\delta},$$

with $\delta > 0$, an infinitesimal positive number. Observe that $i\delta$ must be introduced to the energy in order to have a convergent integral.

(iii) In the general interacting case, the retarded Green function contains a finite imaginary part in the denominator:

$$G_\sigma^R(\mathbf{k}, \omega) = \frac{1}{\omega - \epsilon_{\mathbf{k}} + i/(2\tau)}.$$

Use the residue theorem to to calculate the time-dependent Green function by Fourier transform. How does the Green function behave in the large $t - t'$ limit? Compare with the time dependence of the non-interacting retarded Green function obtained in (a) to give a physical interpretation of τ. Evaluate the spectral density: $\rho_\sigma(\mathbf{k}, \omega) = -\frac{1}{\pi} Im G_\sigma^R(\mathbf{k}, \omega)$, and compare with the non-interacting spectral density, $\rho_\sigma^{(0)}(\mathbf{k}, \omega)$.

(iv) The ground state properties of the system can be obtained from the causal Green function. Calculate the average particle density, $\langle n(\mathbf{x})\rangle$, and kinetic energy, $\langle T(\mathbf{x})\rangle$ in a free Fermi gas using the expressions in terms of the Green function we have obtained in class: $\langle n(\mathbf{x})\rangle = -i(2S + 1)G(\mathbf{x}, 0^-)|_{\mathbf{x}=0}$ and $\langle T(\mathbf{x})\rangle = i(2S + 1)\frac{\hbar^2\nabla^2}{2m}G(\mathbf{x}, 0^-)|_{\mathbf{x}=0}$, where S is the spin of the particles. Check that the results you find are consistent with your solid state physics courses.

2.5 General Analytic Properties of Green Functions in Interacting Systems

We now consider the properties of the causal Green function of a many-particle interacting system. We extract general analytic properties of the Green function without making assumptions about the nature of the particle interactions. We will provide a general relation between the causal Green function and the retarded Green function which is the relevant one to experiments.

The causal Green function of a general interacting fermionic system reads:

$$G_{\alpha\beta}(x, x') = -i\langle T[\Psi_\sigma^\dagger(\mathbf{x})\Psi_\sigma(\mathbf{x})]\rangle$$
$$= -i\{\theta(t - t')\langle\Psi_\alpha(x)\Psi_\beta^\dagger(x')\rangle - \theta(t - t')\langle\Psi_\beta^\dagger(x')\Psi_\alpha(x)\rangle\}.$$

Introducing a complete orthonormal basis, $\{|n\rangle\}$ of the full hamiltonian with undetermined number of particles we have:

$$G_{\alpha\beta}(\mathbf{x}, t; \mathbf{x}', t') = -i\sum_n (\theta(t - t')\langle\Psi_G(N)|e^{iHt}\Psi_\alpha(\mathbf{x})e^{-iHt}|\Psi_n\rangle$$

$$\times\langle\Psi_n|e^{iHt'}\Psi_\beta^\dagger(\mathbf{x}')e^{-iHt'}|\Psi_G(N)\rangle$$

$$-\theta(t' - t)\langle\Psi_G(N)|e^{iHt}\Psi_\beta^\dagger(\mathbf{x}')e^{-iHt}|\Psi_n\rangle\langle\Psi_n|e^{iHt}\Psi_\alpha(\mathbf{x})e^{-iHt}|\Psi_G(N)\rangle)$$

$$= -i\sum_n (\theta(t - t')e^{-i(E_n - E_G(N))(t-t')}\langle\Psi_G(N)|\Psi_\alpha(\mathbf{x})|\Psi_n\rangle\langle\Psi_n|\Psi_\beta^\dagger(\mathbf{x}')|\Psi_G(N)\rangle$$

$$-\theta(t' - t)e^{i(E - E_G(N))(t-t')}\langle\Psi_G(N)|\Psi_\beta^\dagger(\mathbf{x}')|\Psi_n\rangle\langle\Psi_n|\Psi_\alpha(\mathbf{x})|\Psi_G(N)\rangle).$$

Since:

$$N\psi_\beta(\mathbf{x})|\Psi_G(N)\rangle = (N - 1)\psi_\beta(\mathbf{x})|\Psi_G(N)\rangle, \tag{2.27}$$

the matrix elements $\langle\Psi_G(N)|\Psi_\beta^\dagger(\mathbf{x}')|\Psi_n\rangle$ are non-zero only when $|\Psi_n\rangle$ is in the $N - 1$ particle sector. In the presence of traslational symmetry we have:

$$\Psi_\alpha(\mathbf{x}) = \frac{1}{\sqrt{V}}\sum_\mathbf{k} e^{i\mathbf{k}\cdot\mathbf{x}}c_{\mathbf{k}\alpha}, \; \Psi_\alpha(\mathbf{x}) = \frac{1}{\sqrt{V}}\sum_\mathbf{k} e^{-i\mathbf{k}\cdot\mathbf{x}}c_{\mathbf{k}\alpha}^\dagger, \tag{2.28}$$

where V is the volume, $\mathbf{k}$ is the conserved wavevector and $|\psi_n^{N\pm1}\rangle$ can be classified according to the net system momentum. Using the momentum basis we have:

$$
G_{\alpha\beta}(\mathbf{x}, t; \mathbf{x}', t') = -\frac{1}{V} \sum_{\mathbf{k}} e^{i\mathbf{k}\cdot(\mathbf{x}-\mathbf{x}')}
$$

$$
\times (e^{-i(E_n(N+1)-E_G(N))(t-t')}|\langle\Psi_n(N+1)|c^\dagger_{\mathbf{k}\alpha}|\Psi_G(N)\rangle|^2\theta(t-t')e^{-\eta(t-t')}
$$

$$
- e^{i(E_n(N-1)-E_G(N))(t-t')}\langle\Psi_n(N-1)|c_{\mathbf{k}\alpha}|\Psi_G(N)\rangle\theta(t'-t)e^{\eta(t-t')})
$$

$$
= \frac{1}{V} \sum_{\mathbf{k}} e^{-i\mathbf{k}(\mathbf{x}-\mathbf{x}')} G_\alpha(\mathbf{k}, t-t')\delta_{\alpha\beta}. \tag{2.29}
$$

Its Fourier decomposition reads:

$$
G_\sigma(\mathbf{k}, \omega) = \int\limits_{-\infty}^{\infty} d(t-t') e^{i\omega(t-t')} G_\sigma(\mathbf{k}, t-t')
$$

$$
= \sum_n \left(\frac{|\langle\Psi_n(N+1)|c^\dagger_{\mathbf{k}\sigma}|\Psi_G(N)\rangle|^2}{\omega - (E_n(N+1) - E_G(N)) + i\eta} + \frac{|\langle\Psi_n(N-1)|c_{\mathbf{k}\sigma}|\Psi_G(N)\rangle|^2}{\omega + (E_n(N-1) - E_G(N)) - i\eta} \right).
$$

This is the Lehmann representation of the Green function in a spatially and temporal uniform system. What is the meaning of the energy differences in the denominator? The single poles appearing in the Green function are the excitation energies on adding or substracting one particle to the N-particle system. The excitation energies of adding (substracting) a particle from N to $N+1(N-1)$ are:

$$
\omega = E_n(N+1) - E_G(N) = \epsilon_n(N+1) + \mu
$$

$$
\omega = E_G(N) - E_n(N-1) = -\epsilon_n(N-1) + \mu \tag{2.30}
$$

where $\epsilon_n(N \pm 1)$ are the excitation energies in the $N \pm 1$ particle sectors, $\epsilon_n(N \pm 1) = E_n(N \pm 1) - E_0(N \pm 1)$ and $\mu = E_0(N+1) - E_0(N)$ is the chemical potential which is equal to $\mu = E_0(N) - E_0(N-1)$ for sufficiently large N. Note that $\epsilon_n(N \pm 1) \geq 0$ by definition of the excitation energies.

Exercise I.7: Lehmann representation of Green function of the Hubbard dimer

Consider the dimer Hubbard model:

$$
H = -t \sum_\sigma (c^\dagger_{1\sigma}c_{2\sigma} + c^\dagger_{2\sigma}c_{1\sigma}) + U(n_{1\uparrow}n_{1\downarrow} + n_{2\uparrow}n_{2\downarrow}), \tag{2.31}
$$

at half-filling. Obtain the Lehmann representation of the Green function: $G_{ii\sigma}(\omega)$, where $i = 1, 2$ is a site of the dimer for $U = 0$ and $U >> t$. Sketch out the evolution of $G_{ii\sigma}(\omega)$ with U/t interpreting the results found.

2.6   Retarded and Advanced Green Function

These Green functions are defined as:

$$G_{\alpha\beta}^{R}(x, x') = -i \langle \{\Psi_\alpha(x), \Psi_\beta^\dagger(x')\}\rangle \theta(t - t')$$

$$G_{\alpha\beta}^{A}(x, x') = i \langle \{\Psi_\alpha(x), \Psi_\beta^\dagger(x')\}\rangle \theta(t' - t)$$

The retarded Green function is directly related to the intensities in photoemission experiments. Both $G_{\alpha\beta}^{R/A}$ can be related to the time-ordered or causal Green functions as we will see. The Lehmann representation of $G_{\alpha\beta}^{R/A}(\mathbf{k}, \omega)$ can be obtained following the steps for obtaining $G_{\alpha\beta}(\mathbf{k}, \omega)$ and reads:

$$G_\sigma^{R/A}(\mathbf{k}, \omega) = \sum_n \left(\frac{|\langle \Psi_n(N+1)|c_{\mathbf{k}\sigma}^\dagger|\Psi_G(N)\rangle|^2}{\omega - (E_n(N+1) - E_G(N)) \pm i\eta} + \frac{|\langle \Psi_n(N-1)|c_{\mathbf{k}\sigma}|\Psi_G(N)\rangle|^2}{\omega + (E_n(N-1) - E_G(N)) \pm i\eta} \right).$$

$$(2.32)$$

where the $+(-)$ in the denominator corresponds to the R(A) Green functions. In contrast to the causal Green functions, $G_\sigma^{R/A}(\mathbf{k}, \omega)$, are analytic in the upper (R) or the lower (A) imaginary part of the complex ω-plane (Fig. 2.2).

We can express the $G_\sigma^{R/A}(\mathbf{k}, \omega)$ in terms of the spectra densities:

$$G_\sigma^{R/A}(\mathbf{k}, \omega) = \int_{-\infty}^{\infty} \frac{\rho_\sigma(\mathbf{k}, \omega')}{\omega - \omega' \pm i\eta} d\omega' \tag{2.33}$$

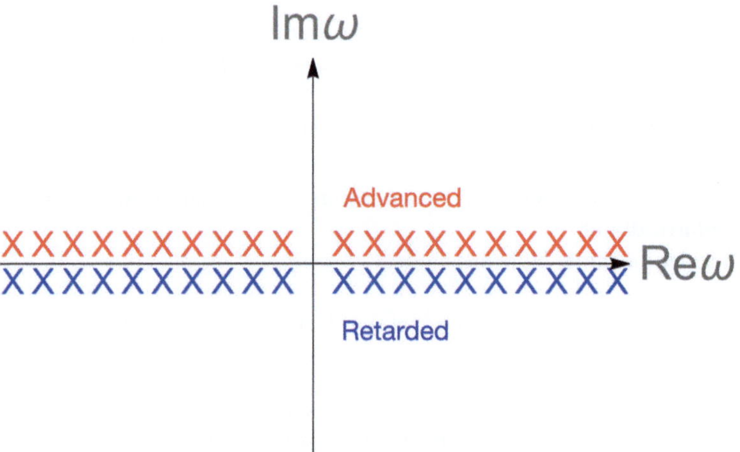

Fig. 2.2 Pole structure of retarded and advanced Greens functions in the complex ω-plane. Excitation energies are referred to the Fermi energy

where we have defined the spectral density: $\rho_\sigma(\mathbf{k}, \omega) = \rho_{e\sigma}(\mathbf{k}, \omega)\theta(\omega - \mu) + \rho_{h\sigma}(\mathbf{k}, \omega)\theta(\mu - \omega)$. So from the above expressions we can obtain the spectral density of states or spectral function:

$$\rho_\sigma(\mathbf{k}, \omega) = \mp \frac{1}{\pi} Im G_\sigma^{R/A}(\mathbf{k}, \omega) = \mp \frac{1}{\pi} Im G_\sigma(\mathbf{k}, \omega + i\delta). \tag{2.34}$$

The spectral density satisfies an important sum rule:

$$\int\limits_{-\infty}^{\infty} \rho_\sigma(\mathbf{k}, \omega) = \int\limits_{-\infty}^{\mu} \rho_{h\sigma}(\mathbf{k}, \omega)d\omega + \int\limits_{\mu}^{\infty} \rho_{e\sigma}(\mathbf{k}, \omega)d\omega$$

$$= \sum_n |\langle \Psi_n(N-1)|c_{\mathbf{k}\sigma}|\Psi_G(N)\rangle|^2 + |\langle \Psi_n(N+1)|c_{\mathbf{k}\sigma}^\dagger|\Psi_G(N)\rangle|^2$$

$$= \langle \Psi_G(N)|c_{\mathbf{k}\sigma}^\dagger c_{\mathbf{k}\sigma} + c_{\mathbf{k}\sigma}c_{\mathbf{k}\sigma}^\dagger|\Psi_G(N)\rangle = 1 \tag{2.35}$$

Hence:

$$\int\limits_{-\infty}^{\infty} \rho_\sigma(\mathbf{k}, \omega) = 1 \tag{2.36}$$

which is always valid independently of the strength of interactions. This sum rule can be used to obtain the asymptotic frequency dependence of the Green functions. In the limit $\omega \to \infty$ we have:

$$G(\mathbf{k}, \omega \to \infty) = G^{R/A}(\mathbf{k}, \omega) \sim \frac{1}{\omega} \int\limits_{-\infty}^{\infty} d\omega' \rho_\sigma(\mathbf{k}, \omega') = \frac{1}{\omega}, \tag{2.37}$$

which is valid independently of the strength of the interactions.

2.7 Fermi Liquid Properties

We now discuss some important general properties of Fermi liquids in terms of the Green function formalism introduced. We first discuss the non-interacting Fermi gas.

As we have seen, the causal or time-ordered Green function of non-interacting electrons reads:

$$G(\mathbf{k}, \omega) = \frac{\theta(k - k_F)}{\omega - \epsilon_\mathbf{k} + i\eta} + \frac{\theta(k_F - k)}{\omega - \epsilon_\mathbf{k} - i\eta}, \tag{2.38}$$

from which we can define the retarded and advanced Green functions:

$$G^{R/A}(\mathbf{k}, \omega) = G(\mathbf{k}, \omega \pm i\eta) = \frac{1}{\omega - \epsilon_\mathbf{k} \pm i\eta}, \tag{2.39}$$

from which we can extract the spectral density:

$$\rho_\sigma(\mathbf{k}, \omega) = \mp Im G_\sigma^{R/A}(\mathbf{k}, \omega) = \delta(\omega - \epsilon_\mathbf{k}). \tag{2.40}$$

where we have used the relation:

$$\frac{1}{\omega \pm i\eta} = \text{P.P.} \int \frac{d\omega}{\omega - \epsilon_\mathbf{k}} \mp i\pi\delta(\omega - \epsilon_\mathbf{k}). \tag{2.41}$$

Thus, the spectral density of non-interacting electrons is a just a Dirac delta function fixed at $\epsilon_\mathbf{k}$. An electron (hole) added (substracted) on (from) a state $\mathbf{k}$ with $|\mathbf{k}| > k_F$ ($|\mathbf{k}| < k_F$) stays indifinetely in that state with corresponding energy $\epsilon_\mathbf{k}$. This is because states in a non-interacting Fermi gas are stationary; no decay processes are possible. Note that the sum rule is automatically satisfied, as it should.

The time-dependence of the R and A Green functions can be obtained from a Fourier transform:

$$G^{R/A}(\mathbf{k}, t) = \int_{-\infty}^{\infty} \frac{d\omega}{2\pi} e^{-i\omega t} G^{R/A}(\mathbf{k}, \omega) = \mp i e^{-i\epsilon_\mathbf{k} t}\theta(\pm t). \tag{2.42}$$

From the causal Green function we can find the occupation number:

$$\langle n_{\mathbf{k}\sigma} \rangle = -iG(\mathbf{k}, 0^-) = -i(i\theta(k_F - |\mathbf{k}|)e^{-i\epsilon_\mathbf{k} 0^-}) = \theta(k_F - |\mathbf{k}|), \tag{2.43}$$

as it should. Alternatively, the occupation may be expressed explicitly in terms of the spectral density:

$$\langle n_{\mathbf{k}\sigma} \rangle = -iG(\mathbf{k}, 0^-) = \int_{-\infty}^{\infty} \frac{d\omega}{2\pi} e^{-i\omega 0^-} G(\mathbf{k}, \omega) = \int_{-\infty}^{\mu} d\omega \rho_\sigma(\mathbf{k}, \omega), \tag{2.44}$$

which is valid for any system. For non-interacting electrons it reduces to:

$$\langle n_{\mathbf{k}\sigma} \rangle = \int_{-\infty}^{0} d\omega \rho_\sigma(\mathbf{k}, \omega) = \int_{-\infty}^{0} d\omega \delta(\omega - \epsilon_\mathbf{k}) = \theta(k_F - |\mathbf{k}|), \tag{2.45}$$

as it should.

What happens when interactions between electrons in the Fermi gas are turned on? Under some general assumptions Landau was able to describe general features of the Fermi liquid (= Fermi gas + Coulomb interactions). A crucial characteristic feature of the Fermi liquid is the existence of quasiparticles (Landau 1957). In the Green function framework, this means that a prominent peak is present in $\rho(\mathbf{k}, \omega)$ which is reminiscent of the Dirac delta peak of non-interacting electrons. Indeed,

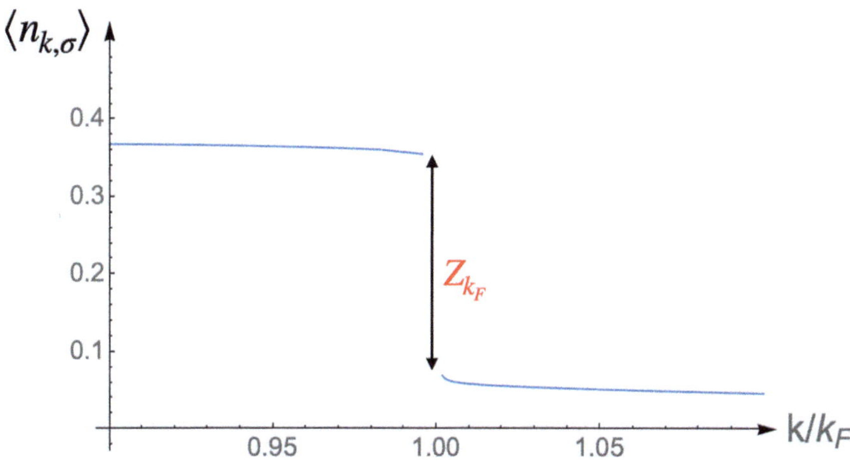

Fig. 2.3 Quasiparticle contribution to the electron occupation number in a Fermi liquid. Even in the presence of the Coulomb interaction, there is a jump in the electron occupation number, $\langle n_{\mathbf{k}\sigma} \rangle$, at the Fermi energy which implies the existence of a Fermi surface. Unlike in a Fermi gas the size of the jump given by the quasiparticle weight, Z_{k_F}, can be smaller than 1: $0 < Z_{k_F} \leq 1$

for electrons at the Fermi surface, $|\mathbf{k}| = k_F$, the spectral function of a Fermi liquid reads:

$$\rho_\sigma(\mathbf{k}, \omega) = Z_{\mathbf{k}} \delta(\omega - \tilde{\epsilon}_{\mathbf{k}}) + \rho_\sigma^{inc.}(\mathbf{k}, \omega). \tag{2.46}$$

where $Z_{\mathbf{k}}$ is the quasiparticle weight or the fraction of the single particle excitation which behaves as a non-interacting electron. The incoherent part of the spectra is denoted by $\rho_{inc.}(\mathbf{k}, \omega)$. From the sum rule for $\rho(\mathbf{k}, \omega)$ we can extract:

$$Z_{\mathbf{k}} = 1 - \int_{-\infty}^{\infty} \rho_\sigma^{inc}(\mathbf{k}, \omega) d\omega, \tag{2.47}$$

which implies: $Z_{\mathbf{k}} \leq 1$.

Hence, the existence of a quasiparticle peak in $\rho(\mathbf{k}, \omega)$ is a signature of the existence of a Fermi liquid, i.e., a gas of weakly interacting quasi-particles. However, this is not the only possible metallic behavior. Metals with no quasiparticle peaks i.e. with featureless $\rho(\mathbf{k}, \omega)$ lie out of the applicability of Landau's theory of the Fermi liquid. In fact such metals are known as non-Fermi liquids. At present there is no systematic understanding of these fascinating metallic states beyond the Fermi liquid paradigm. How does the occupation $\langle n_{\mathbf{k},\sigma} \rangle$ behaves?

Introducing (2.46) in the expression:

$$\langle n_{\mathbf{k}\sigma} \rangle = \int_{-\infty}^{0} d\omega \rho_\sigma(\mathbf{k}, \omega), \tag{2.48}$$

we can infer that $\langle n_{\mathbf{k}\sigma}\rangle$ must have a jump of size $Z_{\mathbf{k}} \leq 1$ at $k = k_F$. This jump in $\langle n_{\mathbf{k}\sigma}\rangle$ is a crucial general property of Fermi liquids. In spite of the Coulomb interaction, the occupation number behaves similarly to the Fermi factor of a non-interacting Fermi gas. However, the jump when crossing the Fermi energy is reduced from 1 due to renormalization effects. In Fig. 2.3 we show the occupation in a Fermi liquid. Note that even at $T = 0$ there is non-zero probability of occupying states above the Fermi energy. This is in contrast to the case of non-interacting electrons.

Thus, Fermi liquids are characterized by the existence of a jump in $\langle n_{\mathbf{k}\sigma}\rangle$, or in other words, the existence of a Fermi surface. In non-Fermi liquids such as the Luttinger liquid, $\langle n_{\mathbf{k}\sigma}\rangle$, does not display a jump (Luttinger 1963). This implies there is no Fermi surface, but since the system is gapless it is still metallic. However, this metal is unconventional in the sense that its elementary excitations are not quasiparticles but they are collective. One can expect that non-Fermi liquids present very different thermodynamic and transport properties than conventional metals described by Landau Fermi liquid theory.

Chapter 3
Perturbation Theory at Zero Temperature

Since, in general, it is difficult to evaluate $G(\mathbf{k}, \omega)$ exactly in an interacting many-partcile system, we introduce the perturbation theory method developed by Feynman and Dyson in particle physics. We generically call this $T = 0$ method, Feynman-Dyson perturbation theory (FDPT). We consider FDPT on the Coulomb interaction and electron-phonon coupling acting on a Fermi gas which is relevant to interacting condensed matter systems. FDPT is much more convenient than the standard perturbation theory used to obtain approximate solutions to Schrödinger's equation.

3.1 Interaction Picture

In order to perform perturbation theory it is convenient to use the interaction picture for operators and states instead of the Heisenberg or Schrödinger representation. Consider the hamiltonian which can be splitted as:

$$H = H_0 + V(t) \tag{3.1}$$

where $V(t)$ is the perturbation we do not know how to treat exactly and can depend explicitly on time. H_0 is the non-interacting part which is time-independent and can be solved exactly. We define a state in the interaction representation as:

$$|\Psi_I(t)\rangle \equiv e^{iH_0 t}|\Psi_s(t)\rangle, \tag{3.2}$$

where $|\Psi_s(t)\rangle$ is the state in the Schrödinger picture. The EOM for $|\Psi_I(t)\rangle$ is:

$$i\frac{\partial}{\partial t}|\Psi_I(t)\rangle = -H_0|\Psi_I(t)\rangle + e^{iH_0 t}(H_0 + V)e^{-iH_0 t}|\Psi_I(t)\rangle, \tag{3.3}$$

J. Merino and A. L. Yeyati, *Many-Body Techniques in Condensed Matter Physics*, UNITEXT for Physics, https://doi.org/10.1007/978-3-031-55143-7_3

which gives:

$$i\frac{\partial}{\partial t}|\Psi_I(t)\rangle = V_I(t)|\Psi_I(t)\rangle. \tag{3.4}$$

where we have used $i\frac{\partial}{\partial t}|\Psi_s(t)\rangle = H|\Psi_s(t)\rangle$ and the definition for the operators in the interaction representation:

$$O_I(t) = e^{iH_0 t}O_s e^{-iH_0 t}, \tag{3.5}$$

which leave the observables invariant: $\langle\Psi_I(t)|O_I(t)|\Psi_I(t)\rangle = \langle\Psi_S(t)|O_S|\Psi_S(t)\rangle$. Note that the equation for $|\Psi_I(t)\rangle$ is valid in general including a time-dependent hamiltonian, $i.e$, for $V = V(t)$.

3.2 Time-Evolution Operator

We now consider the time-evolution operator in the interaction picture:

$$|\Psi_I(t)\rangle = U(t, t')|\Psi_I(t')\rangle. \tag{3.6}$$

which implies $U(t, t) = 1$.

Inserting Eq. (3.6) in Eq. (3.4) one can obtain a general differential equation for $U(t, t')$:

$$i\frac{\partial U(t, t')}{\partial t} = V_I(t)U(t, t'). \tag{3.7}$$

Importantly this equation is valid in general, including the explicitly time-dependent hamiltonian $H = H(t)$!!.

A closed expression for $U(t, t')$ can be obtained only in the case in which H is independent of time. From the definition of states in the Schrödinger representation:

$$|\Psi_S(t)\rangle = e^{-iH(t-t')}|\Psi_S(t')\rangle \to e^{-iH_0 t}|\Psi_I(t)\rangle = e^{-iH(t-t')}e^{-iH_0 t'}|\Psi_I(t')\rangle. \tag{3.8}$$

From which $U(t, t')$ reads:

$$U(t, t') = e^{iH_0 t}e^{-iH(t-t')}e^{-iH_0 t'}. \tag{3.9}$$

We now consider the general solution to $U(t, t')$ from the Eq. (3.7) valid also in the time-dependent case, $H = H(t)$. This solution will prove crucial to the perturbation theory of the Green function. We start by expressing $U(t, t')$ in the integral form:

$$\int_{t'}^{t} dU(t, t') = -i\int_{t'}^{t} V(t_1)U(t_1, t')dt_1. \tag{3.10}$$

Performing the integration on both sides:

$$U(t, t') = 1 - i \int_{t'}^{t} V(t_1)U(t_1, t')dt_1, \tag{3.11}$$

where we have used $U(t', t') = 1$. This equation can be solved iteratively:

$$U(t, t') = 1$$

$$U(t, t') = 1 - i \int_{t'}^{t} V(t_1)(1 - i \int_{t'}^{t} V(t_2)dt_2)dt_1$$

$$= 1 - i \int_{t'}^{t} V(t_1)dt_1 + i^2 \int_{t'}^{t} dt_1 \int_{t'}^{t_1} dt_2 V(t_1)V(t_2)$$

$$U(t, t') = 1 - i \int_{t'}^{t} V(t_1)dt_1 + (-i)^2 \int_{t'}^{t} dt_1 \int_{t'}^{t_1} dt_2 V(t_1)V(t_2) + \cdots$$

$$+ (-i)^n \int_{t'}^{t} dt_1 \int_{t'}^{t_1} dt_2 ... \int_{t'}^{t_{n-1}} dt_n V(t_1)...V(t_n).$$

Note that $t_1 \geq t_2 \geq \cdots \geq t_n$ in the integrations above to the given order n. The expression above can be written in a different form more useful for perturbation series expansions. For instance, the second order term:

$$\int_{t'}^{t} dt_1 \int_{t'}^{t_1} dt_2 V(t_1)V(t_2) = \frac{1}{2}\left[\int_{t'}^{t} dt_1 \int_{t'}^{t_1} dt_2 V(t_1)V(t_2) + \int_{t'}^{t} dt_2 \int_{t_2}^{t} dt_1 V(t_1)V(t_2) \right]$$

$$= \frac{1}{2}\left[\int_{t'}^{t} dt_1 \int_{t'}^{t_1} dt_2 V(t_1)V(t_2) + \int_{t'}^{t} dt_1 \int_{t_1}^{t} dt_2 V(t_2)V(t_1) \right].$$

Summarising:

$$\int_{t'}^{t} dt_1 \int_{t'}^{t_1} dt_2 V(t_1)V(t_2) = \frac{1}{2}\left[\int_{t'}^{t} dt_1 \int_{t'}^{t} dt_2 (V(t_1)V(t_2)\theta(t_1 - t_2) + V(t_2)V(t_1)\theta(t_2 - t_1)) \right]$$

$$= \frac{1}{2}\left[\int_{t'}^{t} dt_1 \int_{t'}^{t} dt_2 T[V(t_1)V(t_2)] \right] \tag{3.12}$$

so the evolution operator can be expressed as a series expansion:

$$U(t, t') = \sum_{n=0}^{\infty} \frac{(-i)^n}{n!} \int_{t'}^{t} dt_1 \dots \int_{t'}^{t} dt_n \, T[V(t_1) \dots V(t_n)], \tag{3.13}$$

or more formally:

$$U(t, t') = T_{exp} \left\{ -i \int_{t'}^{t} dt'' V(t'') \right\}. \tag{3.14}$$

This expression is crucial for the Feynman-Dyson perturbation theory as we will see.

Note that $U(t, t')$ satisfies the following useful properties:

$$U(t, t')U(t', t'') = U(t, t'')$$
$$U(t, t') = U(t', t)^\dagger$$
$$U(t, t')U^\dagger(t, t') = U(t, t')U(t', t) = U(t, t) = 1. \tag{3.15}$$

3.3 Relation Between Heisenberg and Interaction Pictures

Since we have seen that the Green functions are expressed in terms of Heisenberg operators, it is convenient to express these in the interaction picture for the FDPT.

As we already discussed Heisenberg state vectors are independent of time. Since, from the definition of states in the interaction picture $|\Psi_I(0)\rangle = |\Psi_s(0)\rangle$ and $|\Psi_H\rangle = |\Psi_s(0)\rangle$

$$|\Psi_H\rangle = |\Psi_I(0)\rangle. \tag{3.16}$$

The relation between operators can be obtained from the condition that average values of operators should equate in the Heisenberg and interaction representations. Using this condition and the definition of $U(t, t')$ in the interaction representation and $|\Psi_H\rangle = |\Psi_I(0)\rangle$ we find:

$$\langle \Psi_I(t)|O_I(t)|\Psi_I(t)\rangle = \langle \Psi_I(0)|U^\dagger(t, 0)O_I(t)U(t, 0)|\Psi_I(0)\rangle = \langle \Psi_H|O_H(t)|\Psi_H\rangle \tag{3.17}$$

from which:

$$O_H(t) = U^\dagger(t, 0)O_I(t)U(t, 0) \tag{3.18}$$

In the case of a time-independent $H \neq H(t)$, the Heisenberg state vectors can be formally expressed in the interaction picture as:

$$|\Psi_H\rangle = e^{iHt}|\Psi_S(t)\rangle = e^{iHt}e^{-iH_0t}|\Psi_I(t)\rangle, \qquad (3.19)$$

which is completely general for both time-dependent and time independent H. Using this expression, the Heisenberg operators can be explicitly expressed in the interaction picture as:

$$O_H(t) = e^{iHt}O_s e^{-iHt} = e^{iHt}e^{-iH_0t}O_I(t)e^{iH_0t}e^{-iHt} \qquad (3.20)$$

which allows obtaining an explicit expression of U when H is time-independent:

$$U(t,0) = e^{iH_0t}e^{-iHt}. \qquad (3.21)$$

3.4 Adiabatic Hypothesis

Refers to the idea of generating exact eigenstates of interacting system from those of the non-interacting system. In order to be able to perform FDPT on the full Green function of the interacting model described by H we need to express the Green function in terms of the ground state of the non-interacting system, $|\Phi_0\rangle$, described by H_0. Can we somehow relate $|\Psi_G\rangle$ and $|\Phi_G\rangle$?

This can be done through the adiabatic hypothesis which involves certain key assumptions. We assume that in $t = -\infty$ there is no interaction and the system is in the unperturbed state since the interaction is adiabatically turned on through:

$$H = H_0 + V(t) = H_0 + \lim_{\epsilon \to 0^+} V_\epsilon(t) = H_0 + \lim_{\epsilon \to 0^+} e^{-\epsilon|t|}V, \qquad (3.22)$$

where $V_\epsilon(t)$ is shown in Fig. 3.1.

Under such adiabatic switching on of the interaction we can generate eigenstates of H from eigenstates of H_0. So while at $t = \pm\infty$, we recover the non-interacting model, H_0, at $t = 0$ we have the full interacting hamiltonian, H. At the end of the calculations we will take the limit: $\epsilon \to 0^+$. For the results to make sense they should be independent of ϵ.

We have that:

$$|\Psi_I(t)\rangle = U_\epsilon(t, t')|\Psi_I(t')\rangle, \qquad (3.23)$$

which means:

$$|\Psi_I(0)\rangle = U_\epsilon(0, -\infty)|\Psi_I(-\infty)\rangle, \qquad (3.24)$$

where $|\Psi_I(0)\rangle$ is an eigenstate of the full hamiltonian H and $|\Psi_I(-\infty)\rangle$ of the non-interacting hamiltonian H_0 in the interaction representation.

So what are these states $|\Psi_I(0)\rangle$ and $|\Psi_I(-\infty)\rangle$ From:

$$|\Psi_H\rangle = U(0, t)|\Psi_I(t)\rangle \Rightarrow |\Psi_I(0)\rangle = |\Psi_H\rangle, \qquad (3.25)$$

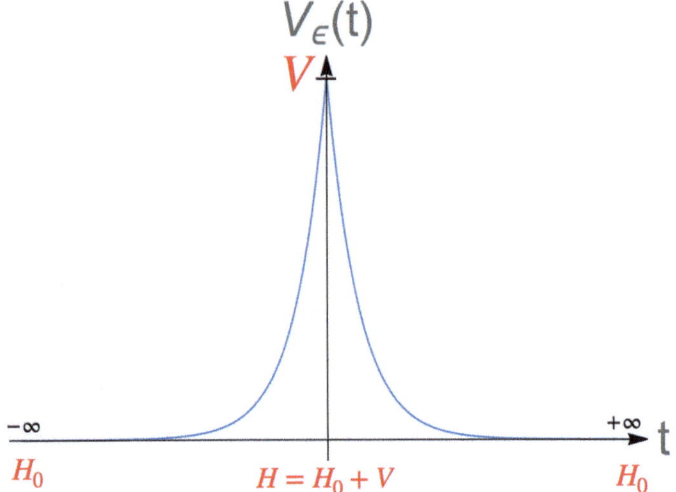

Fig. 3.1 Adiabatic switching on of the interaction. In the limit $t \to \pm\infty$ interactions are suppressed so the system is described by the non-interacting hamiltonian, H_0. At $t = 0$ the full interaction, V, is recovered and the system is described by full hamiltonian, $H = H_0 + V$. By taking $\epsilon \to 0^+$ the interaction is switched on adiabatically

the Heisenberg representation of states coincides with the interaction representation at $t = 0$. Since we also have $|\Psi_H\rangle = |\Psi_S(0)\rangle$ from the definition of Heisenberg state we conclude that $|\Psi_I(0)\rangle = |\Psi_H\rangle = \Psi_S(0)\rangle$ is just an exact eigenstate of the full hamiltonian. Thus:

$$H|\Psi_H\rangle = E|\Psi_H\rangle. \tag{3.26}$$

On the other hand, $|\Psi_I(-\infty)\rangle$, satisfies:

$$\lim_{t_0 \to -\infty} |\Psi_I(t_0)\rangle = e^{i H_0 t_0}|\Psi_s(t_0)\rangle \tag{3.27}$$

from its definition. Since we also have that $|\Psi_s(t_0)\rangle$ is a stationary state of the non-interacting hamiltonian H_0, i. e. $|\Psi_s(t_0)\rangle = e^{-i H_0 t_0}|\Psi_0\rangle$, where $|\Psi_0\rangle$ is some eigenstate of the H_0: $H_0|\Phi_0\rangle = E_0|\Phi_0\rangle$ which leads to:

$$|\Psi_I(-\infty)\rangle = |\Psi_0\rangle. \tag{3.28}$$

Hence, we conclude:

$$|\Psi_H\rangle = U_\epsilon(0, -\infty)|\Psi_0\rangle \tag{3.29}$$

This means that we can generate an eigenstate of the full interacting hamiltonian H from the some eigenstate of H_0 by adiabatically switching on the interaction.

But what happens when we take the $\epsilon \to 0$ limit, is it well defined? The answer to this question was sorted out by Gell-Mann and Low.

3.5 Gell-Mann and Low Theorem

The theorem states that if the limit:

$$\lim_{\epsilon \to 0^+} \frac{U_\epsilon(0, -\infty)|\Phi_0\rangle}{\langle\Phi_0|U_\epsilon(0, -\infty)|\Psi_0\rangle} \equiv \frac{|\Psi_0\rangle}{\langle\Phi_0|\Psi_0\rangle}, \tag{3.30}$$

exists to all orders in perturbation theory the above quantity satisfies:

$$H\frac{|\Psi_0\rangle}{\langle\Phi_0|\Psi_0\rangle} = E\frac{|\Psi_0\rangle}{\langle\Phi_0|\Psi_0\rangle}. \tag{3.31}$$

So $\frac{|\Psi_0\rangle}{\langle\Phi_0|\Psi_0\rangle}$ is an eigenstate of H!!.

The state $\frac{|\Psi_0\rangle}{\langle\Phi_0|\Psi_0\rangle}$ should include the denominator in order for the limit to exist. Taken separately the numerator and denominator the limit does not exist since $\lim_{\epsilon\to 0^+} U_\epsilon(0, -\infty)|\Phi_0\rangle.$ does no exist since the phase of this quantity diverges as $1/\epsilon$ but the full quantity exists due to cancellation of numerator and denominator. The state $\frac{|\Psi_0\rangle}{\langle\Phi_0|\Psi_0\rangle}$ does not have to be the ground state of the interacting system, but it is only an eigenstate. The adiabatic hypothesis allow us to assume additionally that the states of the interacting system evolve from the non-interacting system without level crossings. The adiabatic hypothesis then excludes instabilities such as magnetism or superconductivity i.e. no spontaneous broken symmetry solutions are allowed. Hence, based on Gell-Mann and Low theorem and the adiabatic hypothesis we can assume that the ground state of the interacting system:

$$\frac{|\Psi_G\rangle}{\langle\Phi_G|\Psi_G\rangle} = \lim_{\epsilon \to 0^+} \frac{U_\epsilon(0, -\infty)|\Phi_G\rangle}{\langle\Phi_G|U_\epsilon(0, -\infty)|\Psi_G\rangle}. \tag{3.32}$$

Based on all this we can obtain an expression of the Green function in terms of the non-interacting ground state of H_0, $|\Phi_G\rangle$ instead of the ground state of H, $|\Psi_G\rangle$.

We start from the expression:

$$|\Psi_G\rangle = U_\epsilon(0, -\infty)|\Phi_G\rangle = U_\epsilon(0, +\infty)U_\epsilon(+\infty, -\infty)|\Phi_G\rangle = U_\epsilon(0, +\infty)S|\Phi_G\rangle \tag{3.33}$$

where we have introduced the S-matrix, $S = U_\epsilon(+\infty, -\infty)$. Since the interaction is zero for $t = +\infty$:

$$U_\epsilon(\infty, -\infty)|\Phi_G\rangle = S|\Phi_G\rangle = e^{i\alpha}|\Phi_G\rangle. \tag{3.34}$$

since the state at $t = \infty$ and $t = -\infty$ can only differ in a phase.

For $t > 0$ we have:

$$
\begin{aligned}
G(\mathbf{x}, t) &= -i\langle\Psi_G|\Psi(\mathbf{x}, t)\Psi^\dagger(0, 0)|\Psi_G\rangle \\
&= -i\langle\Phi_G|S^\dagger U_\epsilon^\dagger(0, \infty)U_\epsilon^\dagger(t, 0)\Psi_I(\mathbf{x}, t)U_\epsilon(t, 0)\Psi_I^\dagger(0, 0)U_\epsilon(0, -\infty)|\Phi_G\rangle \\
&= -ie^{-i\alpha}\langle\Phi_G|U_\epsilon(\infty, t)\Psi_I(\mathbf{x}, t)U_\epsilon(t, 0)\Psi_I^\dagger(0, 0)U_\epsilon(0, -\infty)|\Phi_G\rangle \\
&= -ie^{-i\alpha}\langle\Phi_G|T[\Psi_I(\mathbf{x}, t)\Psi_I^\dagger(0, 0)U_\epsilon(\infty, t)U_\epsilon(t, 0)U_\epsilon(0, -\infty)]|\Phi_G\rangle
\end{aligned}
$$

Since $\langle\Phi_G|S|\Phi_G\rangle = e^{i\alpha}$:

$$
G(\mathbf{x}, t) = -i\frac{\langle\Phi_G|T[\Psi_I(\mathbf{x}, t)\Psi_I^\dagger(0, 0)S]|\Phi_G\rangle}{\langle\Phi_G|S|\Phi_G\rangle} \tag{3.35}
$$

This expression is the starting point for FDPT. We perform perturbation theory on the $S = U(\infty, -\infty)$ matrix since this is the only quantity where the interaction, V, enters.

Introducing the series expansion for $S = U(\infty, -\infty)$ in $G(\mathbf{x}, t)$ we have:

$$
\begin{aligned}
G(\mathbf{x}, t) &\propto G^{(0)}(\mathbf{x}, t) \\
&= -i\sum_{n=1}^{\infty}\int_{-\infty}^{\infty} dt_1 \int_{-\infty}^{\infty} dt_2... \int_{-\infty}^{\infty} dt_n \langle\Phi_G|T[\Psi(\mathbf{x}, t)\Psi^\dagger(0, 0)V(t_1)V(t_2)...V(t_n)]|\Phi_G\rangle
\end{aligned}
$$

where we have not considered, for the time being the denominator, $\langle\Phi_G|S|\Phi_G\rangle$ entering the definition of the Green function. In spite of having been able to express the Green function in terms of the non-interacting ground state we still need to evaluate expectation values involving a large number of Ψ, $\Psi^\dagger$ field operators. Wick's theorem provides a systematic way to evaluate them.

3.6 Wick's Theorem

Since the interaction contains products of fermionic operators such as the Coulomb interaction: $V \sim \Psi^\dagger\Psi^\dagger\Psi\Psi$ we need a way to evaluate mean values of the type:

$$
\langle\Phi_G|\Psi(1)\Psi(2)...\Psi(n)\Psi^\dagger(n')...\Psi^\dagger(2')\Psi^\dagger(1')|\Phi_G\rangle \tag{3.36}
$$

where $|\Phi_G\rangle$ is the ground state of the non-interacting fermionic system. We are using the compact notation $(i) \rightarrow (\mathbf{x}_i, t_i)$, and the operators are given in the interaction representation:

$$\Psi(\mathbf{x}, t) = e^{iH_0 t}\Psi(\mathbf{x})e^{-iH_0 t}$$

$$i\frac{\partial}{\partial t}\Psi(\mathbf{x}, t) = [\Psi(\mathbf{x}, t), H_0] \tag{3.37}$$

where H_0 must be a quadratic hamiltonian in the fermionic field operators.

Wick's theorem allows to calculate such averages as follows:

$$\langle\Phi_G|T[\Psi(1)\Psi(2)...\Psi(n)\Psi^{\dagger}(n')...\Psi^{\dagger}(2')\Psi^{\dagger}(1')]|\Phi_G\rangle =$$

$$\pm \quad \langle\Phi_G|T[\Psi(1)\Psi^{\dagger}(1')]|\Phi_G\rangle\langle\Phi_G|T[\Psi(2)\Psi^{\dagger}(2')]\Phi_G\rangle \cdots \langle\Phi_G|T[\Psi(n)\Psi^{\dagger}(n')]\Phi_G\rangle$$

$$\pm \quad \langle\Phi_G|T[\Psi(1)\Psi^{\dagger}(2')]|\Phi_G\rangle\langle\Phi_G|T[\Psi(2)\Psi^{\dagger}(1')]\Phi_G\rangle \cdots \pm$$

$$= \quad \text{all possible contractions of the operators } \Psi \text{ and } \Psi^{\dagger} \text{ multiplied by } (-1)^P,$$

where P is the parity of the permutation needed to bring the operators to the original order. For example, we have:

$$\langle\Phi_G|T[\Psi(1)\Psi(2)\Psi^{\dagger}(2')\Psi^{\dagger}(1')]|\Phi_G\rangle =$$

$$- \langle\Phi_G|T[\Psi(1)\Psi^{\dagger}(2')]|\Phi_G\rangle\langle\Phi_G|\Psi(2)\Psi^{\dagger}(1')|\Phi_G\rangle$$

$$+ \langle\Phi_G|T[\Psi(1)\Psi^{\dagger}(1')]|\Phi_G\rangle\langle\Phi_G|\Psi(2)\Psi^{\dagger}(2')|\Phi_G\rangle \tag{3.38}$$

3.7 Diagrammatic Approach in Coordinate Space

From Wick's theorem we have seen how we are able to convert each term in the expansion of the Green function in a product of single particle Green function of the non-interacting (or quadratic) hamiltonian H_0. Lets consider a Coulomb interaction of the form:

$$V = \frac{1}{2}\int \Psi_{\alpha}^{\dagger}(\mathbf{x}_1)\Psi_{\beta}^{\dagger}(\mathbf{x}_2)V(\mathbf{x}_1 - \mathbf{x}_2)\Psi_{\beta}^{\dagger}(\mathbf{x}_2)\Psi_{\alpha}^{\dagger}(\mathbf{x}_1)d\mathbf{x}_1 d\mathbf{x}_2. \tag{3.39}$$

The Coulomb interaction in condensed matter can be assumed to be instantaneous (in contrast to particle physics):

$$V(x_1 - x_2) = V(\mathbf{x}_1 - \mathbf{x}_2)\delta(t_1 - t_2) \tag{3.40}$$

which can then be expressed in terms of four-dimensional integrals as:

$$\int dt1\, V(t_1) = \frac{1}{2}\int d^4x_1 \int d^4x_2 \Psi_{\alpha}^{\dagger}(x_1)\Psi_{\beta}^{\dagger}(x_2)V(x_1 - x_2)\Psi_{\beta}(x_2)\Psi_{\alpha}(x_1) \tag{3.41}$$

and we can treat space and time in an equivalent way in the perturbation expansion. Note that the time dependence implicit in $\Psi(x)$, $\Psi^\dagger(x)$ correspond to the interaction picture.

Lets now analyze the first order terms arising in the expansion of $G(x, x')$:

$$G_{\alpha\beta}(x, x') = G^{(0)}_{\alpha\beta}(x, x') + \delta G^{(1)}_{\alpha\beta}(x, x') + \delta G^{(2)}_{\alpha\beta}(x, x') + \cdots + \delta G^{(n)}_{\alpha\beta}(x, x')$$

(3.42)

The interesting $n = 1$ correction term reads:

$$\begin{aligned}
\delta G^{(1)}_{\alpha\beta}(x, x') &= \frac{(-i)^2}{\langle S\rangle} \int dt_1 \langle T[\Psi_\alpha(x)\Psi^\dagger_\beta(x')V(t_1)]\rangle \\
&= -\frac{1}{\langle S\rangle} \int d^4x_1 \int d^4x_2 \\
&\quad \langle \Phi_G|T[\Psi_\alpha(x)\Psi^\dagger_\beta(x')\Psi^\dagger_\gamma(x_1)\Psi^\dagger_\delta(x_2)\Psi_\delta(x_2)\Psi_\gamma(x_1)]|\Phi_G\rangle \frac{V(x_1 - x_2)}{2},
\end{aligned}$$

which can be decoupled applying Wick's theorem to the average over the product of the six fermion field operators:

$$\begin{aligned}
&\langle T[\Psi_\alpha(x)\Psi^\dagger_\beta(x')\Psi^\dagger_\gamma(x_1)\Psi^\dagger_\delta(x_2)\Psi_\delta(x_2)\Psi_\gamma(x_1)]\rangle = \\
&+ \langle T[\Psi_\alpha(x)\Psi^\dagger_\gamma(x_1)]\rangle \langle \Psi^\dagger_\delta(x_2)\Psi_\delta(x_2)\langle T[\Psi_\gamma(x_1)\Psi^\dagger_\beta(x')\rangle \\
&- \langle T[\Psi_\alpha(x)\Psi^\dagger_\gamma(x_1)]\rangle \langle \Psi^\dagger_\delta(x_2)\Psi_\gamma(x_1)\langle T[\Psi_\delta(x_2)\Psi^\dagger_\beta(x')]\rangle \\
&+ \langle T[\Psi_\alpha(x)\Psi^\dagger_\delta(x_2)]\rangle \langle \Psi^\dagger_\gamma(x_1)\Psi_\gamma(x_1)\langle T[\Psi_\delta(x_2)\Psi^\dagger_\beta(x')]\rangle \\
&- \langle T[\Psi_\alpha(x)\Psi^\dagger_\delta(x_2)]\rangle \langle \Psi^\dagger_\gamma(x_1)\Psi_\delta(x_2)\langle T[\Psi_\gamma(x_1)\Psi^\dagger_\beta(x')]\rangle \\
&+ \langle T[\Psi_\alpha(x)\Psi^\dagger_\beta(x')]\rangle \langle \Psi^\dagger_\gamma(x_1)\Psi_\gamma(x_1)\langle T[\Psi^\dagger_\delta(x_2)\Psi^\dagger_\delta(x_2)]\rangle \\
&- \langle T[\Psi_\alpha(x)\Psi^\dagger_\beta(x')]\rangle \langle \Psi^\dagger_\gamma(x_1)\Psi_\delta(x_2)\langle T[\Psi^\dagger_\delta(x_2)\Psi^\dagger_\gamma(x_1)]\rangle \\
&= i G^{(0)}_{\alpha\gamma}(x, x_1)G^{(0)}_{\delta\delta}(x_2, x_2)G^{(0)}_{\gamma\beta}(x_1, x') - i G^{(0)}_{\alpha\gamma}(x, x_1)G^{(0)}_{\delta\gamma}(x_1, x_2)G^{(0)}_{\delta\beta}(x_2, x') \\
&+ i G^{(0)}_{\alpha\delta}(x, x_2)G^{(0)}_{\gamma\gamma}(x_1, x_1)G^{(0)}_{\delta\beta}(x_2, x') - i G^{(0)}_{\alpha\delta}(x, x_2)G^{(0)}_{\delta\gamma}(x_2, x_1)G^{(0)}_{\gamma\beta}(x_1, x') \\
&- i G^{(0)}_{\alpha\beta}(x, x')G^{(0)}_{\gamma\gamma}(x_1, x_1)G^{(0)}_{\delta\delta}(x_2, x_2) + i G^{(0)}_{\alpha\beta}(x, x')G^{(0)}_{\delta\gamma}(x_2, x_1)G^{(0)}_{\gamma\delta}(x_1, x_2),
\end{aligned}$$

where, for simplicity, we have used the notation, $\langle \Phi_G||\Phi_G\rangle = \langle\rangle$. Note that the equal time Green functions is defined as: $G^{(0)}_{\delta\delta}(x_1, x_2 \equiv G^{(0)}_{\delta\delta}(\mathbf{x}_1 t_1, \mathbf{x}_2 t_2 = t_1 + 0^+)$.

We can conveniently use Feynman diagrams to evaluate each term:

(i) Represent x_1, x_2, x, x' in the plane as points.

(ii) Associate straight solid lines x x_1 to $G^0(x, x_1)$ and wavy lines x_1 x_2 to $V(x_1, x_2)$.

Using the above rules rules, the six terms occurring in $\delta G^{(1)}(x, x')$ above can be drawn as six diagrams:

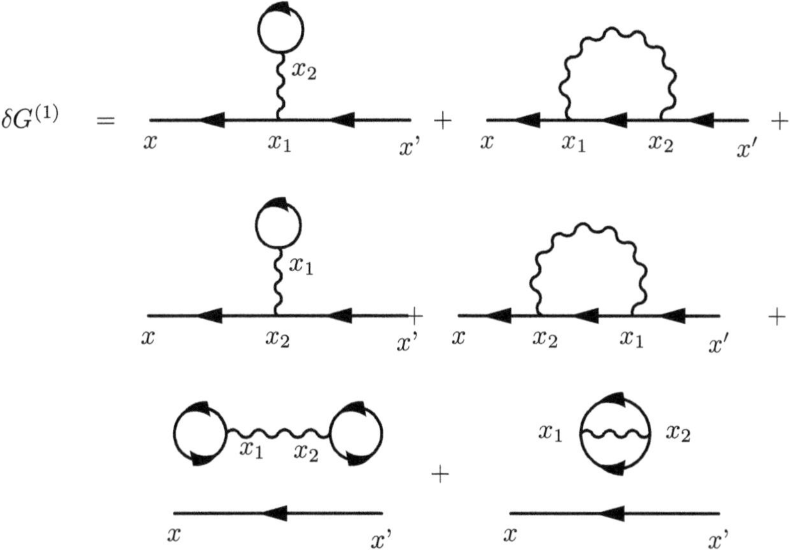

Note that each diagram for $\delta G^{(1)}(x, x')$ has two external x, x' and two internal x_1, x_2 points. Integration over internal points and sum over spin variables is performed. In a similar way, we can associate Feynman diagrams with the mathematical expressions occurring in the expansion of $G(x, x')$ to any order. Hence, we can evaluate $G(x, x')$ to a given order, $\delta G^{(n)}(x, x')$, by forming all Feynman diagrams and calculating the associated mathematical expressions. A set of rules are introduced which associate mathematical formulas with the Feynman diagrams. These Feynman rules greatly simplify the calculations.

There are two type of Feynman diagrams occurring in the expansion of $G(x, x')$: connected and disconnected. These can be recognized in the diagrams for $\delta G^{(1)}$ (x, x'). At any order of the expansion, connected diagrams are those in which $\Psi(x)$ is paired with a $\Psi^\dagger$ in $V(t_1)$, a Ψ in $V(t_1)$ with a $\Psi^\dagger$ in $V(t_2)$,..., until we arrive to $\Psi^\dagger(x')$ without omitting a single $V(t_i)$, The diagrams in which one or several operators $V(t_i)$ are not connected to $\Psi(x)$ and $\Psi^\dagger(x')$ are disconnected.

We now discuss the correction to $G(x, x')$ due to a generic disconnected diagram of order n. This can be expressed as:

$$-i\frac{(-i)^n}{n!}\int dt_1 \ldots dt_m \langle T[\Psi(x)\Psi^\dagger(x')V(t_1)V(t_2)\ldots V(t_m)]\rangle_{conn.}$$

$$\times \int dt_{m+1} \ldots dt_n \langle T[V(t_{\{m+1\}})\ldots V(t_n)]\rangle,$$

where a specific Wick decomposition of the connected and disconnected parts is implicitly assumed since we are, to start with, considering one specific contribution. We notice that among all the possible Wick contractions that can be constructed some of them would give exactly the same answer. For instance, if we exchange the $V(t_i)$ in $\langle\ldots\rangle_{conn}$ with the ones in $\langle\ldots\rangle_{conn}$ this just corresponds to rename the integration variables without modifying the final contribution of the diagram. The number of equivalent diagrams of this type would then be:

$$\frac{n!}{m!(n-m)!} \tag{3.43}$$

so the total contribution of all the diagrams of this type is:

$$(-i)\frac{(-i)^m}{m!}\int dt_1\ldots dt_m \langle T[\Psi(x)\Psi^\dagger(x')V(t_1)\ldots V(t_m)]\rangle_{conn}$$
$$\times\frac{(-i)^{n-m}}{(n-m)!}\int dt_{m+1}\ldots dt_n \langle T[V(t_{m+1})\ldots V(t_n)]\rangle. \tag{3.44}$$

We can now sum all contributions from all the diagrams to any order that contain a common connected part but different/arbitrary disconnected parts with any number $n-m$ de V's:

$$(-i)\frac{(-i)^m}{m!}\int dt_1\ldots dt_m \langle T[\Psi(x)\Psi^\dagger(x')V(t_1)\ldots V(t_m)]\rangle_{conn.}$$
$$\times(1-i\int_{-\infty}^{\infty} dt_{m+1}\langle V(t_{m+1})\rangle - \frac{1}{2}\int_{-\infty}^{\infty} dt_{m+1}\int_{-\infty}^{\infty} dt_{m+2}\langle T[V(t_{m+1})V(t_{m+2})+\cdots)$$

Since the infinite series in the parenthesis is just, $\langle S\rangle = \langle U(\infty,-\infty)\rangle$ we have found that:

$$\langle T[\Psi(x)\Psi^\dagger(x')]\rangle = \langle T[\Psi(x)\Psi^\dagger(x')S]\rangle_{conn.}\langle S\rangle, \tag{3.45}$$

so that the Green function simplifies to:

$$G(x,x') = -i\frac{\langle T[\Psi(x)\Psi^\dagger(x')S]\rangle}{\langle S\rangle} = -i\langle T[\Psi(x)\Psi^\dagger(x')S]\rangle_{conn.} \tag{3.46}$$

so we can omit $\langle S\rangle$ from the denominator and consider only connected diagrams in the evaluation of $G(x,x')$.

Another simplification arises from the conservation that all the connected diagrams arising in the expression:

$$-i\frac{(-i)^m}{m!}\int dt_1\ldots dt_m \langle T[\Psi(x)\Psi^\dagger(x')V(t_1)\ldots V(t_m)]\rangle_{conn.}, \tag{3.47}$$

are identical when permuting the $V(t_i)$. Hence, we can consider only those Wick contractions which lead to topologically inequivalent diagrams i.e. we only consider those diagrams that cannot be obtained from others by a simple permutation of the $V(t_i)$ operators. We will then construct topologically inequivalent diagrams eliminating the $m!$ factor in the denominator.

Some further considerations:

(1) Diagrams which only differ on an interchange of the internal vertices in a given Coulomb interaction give the same contribution since they are topologically equivalent. Such duplication of diagrams can be seen in the Hartree and Fock contributions to $\delta G^{(1)}$ in which x_1 and x_2 interchanged. We will then omit the factor $1/2$ in $V(t_i)$ and just keep the topologically inequivalent diagrams, for instance, only one Hartree and one Fock diagram at order $n = 1$. The simplification done here has to do to interchanges of the vertices of the Coulomb interaction in a particular $V(t_i)$ in contrast to the interchanges between different $V(t_i)$

(2) The overall sign of a diagram has to do with permutations of the field operators. This is connected with the number of closed loops in the diagrams. Hence, the overall sign is, $(-1)^F$, where F is the number of closed loops in a diagram.

(3) We typically encounter contributions containing Green functions evaluated at the same time $G^{(0)}(x_1, x_2)$. This can only occur when two field operators entering $V(t_i)$ are paired up since the Coulomb interaction is instantaneous. These Green's functions should be interpreted as:

$$\lim_{\delta \to 0} G^{(0)}(\mathbf{x}_1 t_1, \mathbf{x}_2, t_2 = t_1 + \delta) = i\langle \Psi^\dagger(\mathbf{x}_2)\Psi(\mathbf{x}_1)\rangle. \tag{3.48}$$

The complete analysis given above can be summarized in a set of rules for the construction of diagrams.

Feynman rules in coordinate space The nth order correction to the Green's function in coordinate space can be obtained following the list of rules:

1. Construct all the topologically inequivalent diagrams with $2n$ internal vertices, x_i, two external vertices (x and x'), $2n + 1$ solid lines and n wavy lines connecting internal vertices.
2. Associate a Green's function $G^{(0)}_{\alpha\beta}(x, x')$ with each solid line,
3. Associate a Coulomb interaction $V(x_1 - x_2) = U(\mathbf{x}_1 - \mathbf{x}_2)\delta(t - t')$ with each wavy line.
4. Integrate over all internal vertices ($d^4 x_i = d\mathbf{x}_i dt_i$) and sum over the internal spin indices.
5. Multiply the resulting expression by the factor: $i^n(-1)^F$, where F is the number of closed fermion loops in the diagram.
6. Green functions like, $G^{(0)}(x_1, x_2)$, containing equal temporal arguments they should be interpreted as:

$$G^{(0)}(x_1, x_2) = \lim_{t \to 0^+} G^{(0)}(\mathbf{x}_1 - \mathbf{x}_2, -t). \tag{3.49}$$

3.8 Diagrammatic Approach in Momentum Space

In spatially uniform and homegeneous in time, the Green function depends on the difference in the two variables, $G(x, x') = G(x - x')$ and perturbation theory greatly simplifies. We illustrate this by considering the first order diagrams in coordinate space introduced previously.

Applying the Feynman rules, the contribution of the Hartree diagram reads:

$$i(-1) \int d^4 x_1 \int d^4 x_2 G^{(0)}_{\alpha\gamma_1}(x, x_1) G^{(0)}_{\gamma_1\beta}(x_1, x') G^{(0)}_{\gamma_2,\gamma_2}(x_2, x_2) V(x_1 - x_2) \quad (3.50)$$

and the exchange or Fock diagram:

$$i(-1)^0 \int d^4 x_1 \int d^4 x_2 G^{(0)}_{\alpha\gamma_1}(x, x_1) G^{(0)}_{\gamma_1\gamma_2}(x_1, x_2) G^{(0)}_{\gamma_2,\beta)}(x_2, x') V(x_1 - x_2) \quad (3.51)$$

Fourier transforming the $G^{(0)}$ and V:

$$G^{(0)}_{\alpha\gamma_1}(x, x_1) = G^{(0)}_{\alpha\gamma_1}(x - x_1) = \int \frac{d^4 k}{(2\pi)^4} G^{(0)}_{\alpha\gamma_1}(k) e^{ik(x-x_1)}$$

$$V(x_1 - x_2) = \int \frac{d^4 q}{(2\pi)^4} V(q) e^{iq(x_1-x_2)} \quad (3.52)$$

where we have used the notation: $p = (\mathbf{p}, \omega)$. Introducing these in the Fock contribution we have:

$$\frac{i}{(2\pi)^{16}} \int d^4 p\, d^4 p_2 d^4 p' d^4 q\, d^4 x_1 d^4 x_2 G^{(0)}_{\alpha\gamma_1}(p) G^{(0)}_{\gamma_1\gamma_2}(p_2) G^{(0)}_{\gamma_2\beta}(p') V(q)$$
$$\times \quad e^{ip(x-x_1)} e^{ip_2(x_1-x_2)} e^{ip'(x_2-x')} e^{iq(x_1-x_2)}$$

Integrating with respect to x_1 and x_2 using:

$$\int e^{-ipx} d^4 x = (2\pi)^4 \delta(p) \quad (3.53)$$

we find:

$$\frac{i}{(2\pi)^8} \int d^4 p\, d^4 p_2 d^4 p' d^4 q\, G^{(0)}_{\alpha\gamma_1}(p) G^{(0)}_{\gamma_1\gamma_2}(p_2) G^{(0)}_{\gamma_2\beta}(p')$$
$$V(q) \delta(p - p_2 - q) \delta(p_2 + q - p') e^{(ipx - ip'x')}$$

Taking the Fourier components with respect to x and x' we find:

$$\delta G^{(1)}_{\alpha\beta}(p, p') = i \int d^4 p_2 \int d^4 q \, G^{(0)}_{\alpha\gamma_1}(p) G^{(0)}_{\gamma_1\gamma_2}(p_2) G^{(0)}_{\gamma_2\beta}(p')$$

$$V(q)\delta(p - p_2 - q)\delta(p_2 + q - p'),$$

where the Dirac δ-functions impose momentum conservation at the internal vertices. Carrying the integration with respect to p_2 and using $G_{\alpha\beta}(p) = G^{(0)}(p)\delta_{\alpha\beta}$ we find:

$$\delta G^{(1)}_{\alpha\beta}(p, p') = i G^{(0)}_{\alpha\gamma_1}(p) \int \frac{d^4 q}{(2\pi)^4} G^{(0)}_{\gamma_1\gamma_2}(p - q) G^{(0)}_{\gamma_2\beta}(p') V(q)(2\pi)^4 \delta(p - p')\delta_{\alpha\gamma_1}\delta_{\gamma_1\gamma_2}\delta_{\gamma_2\beta}$$

$$= \delta G^{(1)}(p)(2\pi)^4 \delta(p - p')\delta_{\alpha\beta}$$

This first order correction, $\delta G^{(1)}(p)$, has an associated Feynman diagram:

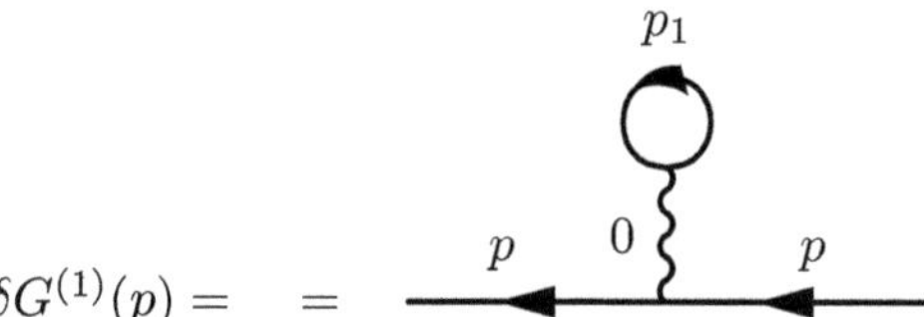

$$\delta G^{(1)}(p) =$$

Performing a similar approach, the Hartree contribution would read:

$$\delta G^{(1)}_{\alpha\beta}(p, p') = -i \int G^{(0)}(p)\delta_{\alpha\beta} V(0) \frac{d^4 p_1}{(2\pi)^4} G^{(0)}(p_1) G^{(0)}(p') e^{i\omega_1 0^+} G^{(0)}(p')(2\pi)^4 \delta(p - p')$$

$$= \delta G^{(1)}(p)(2\pi)^4 \delta(p - p')\delta_{\alpha\beta} \tag{3.54}$$

with corresponding Feynman diagram:

$$\delta G^{(1)}(p) = \quad = \quad$$

where the wavy line in this case carries no energy or momentum $q = (0, 0)$ to conserve momentum in the internal vertices.

Feynman rules in momentum space
The nth order correction in V to the Green's function in momentum space can be obtained from the following rules:

1. Construct all the topologically inequivalent diagrams with $2n$ internal vertices, $2n + 1$ solid lines and n wavy lines.
2. With each line associate a four-component momentum in such a way that the two external lines carry the external momentum and the lines of the internal momentum satisfy energy-momentum conservation at each vertex.
3. Associate with each solid line a Green's function:

$$G^{(0)}_{\alpha\beta}(p) = \frac{\delta_{\alpha\beta}}{\omega + \mu - \epsilon^0_{\mathbf{p}} + i\eta\,\mathrm{sign}(\epsilon^0_{\mathbf{p}} - \mu)} \tag{3.55}$$

4. Associate a Coulomb interaction $V(q)$ with each wavy line.
5. Integrate over all internal four-momenta and sum over the internal spin indices.
6. Multiply the resulting expression by the factor: $i^n(-1)^F(2\pi)^{-4n}$, where F is the number of closed fermion loops in the diagram.
7. Any closed solid line should be interpreted as:

$$G^{(0)}(p_1)e^{i\omega_1 0^+} \tag{3.56}$$

3.9   Self-energies and Dyson Equation

A remarkable property of the diagrammatic theory is that it allows evaluating an infinite set of terms in a perturbation series through the Dyson equation. This can be readily illustrated based on the expansion of the Green function of the interacting electron gas. For example, the diagrammatic expansion up to second order (keeping only some of the terms) reads:

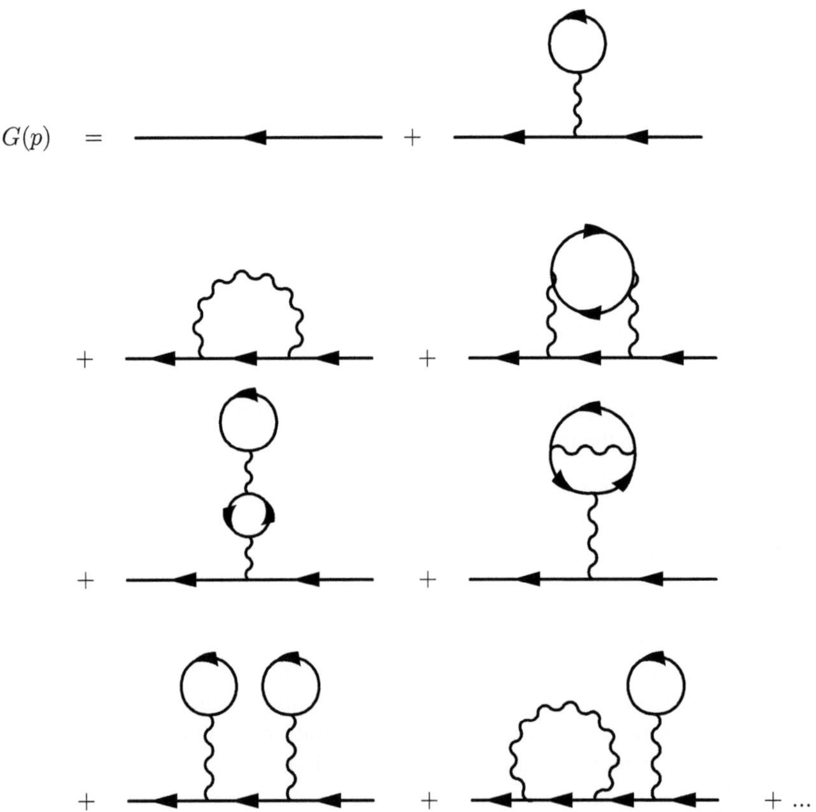

We can sum the infinite perturbation series for $G(p)$ formally introducing an irreducible self-energy, $\Sigma(p)$. A self-energy diagram can be obtained separating from the diagrams for G shown above the external legs. Self-energy diagrams that cannot be splitted into two parts by cutting a single Green function line such as:

are irreducible, whereas diagrams such as:

are examples of reducible self-energies since they can be splitted in two by cutting the middle Green function solid line.

Grouping the irreducible diagrams for the self-energy, $\Sigma(p)$, we have:

$$\Sigma(p) \;=\; \quad + \quad + \quad$$

$+\ldots$

We can generate all diagrams to any order in the perturbation expansion that contribute to $G(p)$ by iterating the equation:

$$G(p) = G^{(0)}(p) + G^{(0)}(p)\Sigma(p)G(p), \tag{3.57}$$

where $\Sigma(p)$ refers to the irreducible part of the full self-energy. Indeed, by iteration we generate the series:

$$\begin{aligned}
G(p) &= G^{(0)}(p) + G^{(0)}(p)\Sigma(p)G^{(0)}(p) + G^{(0)}(p)\Sigma(p)G^{(0)}(p)\Sigma(p)G^{(0)}(p) + \cdots \\
&= G^{(0)}(p)(1 + \Sigma(p)G^{(0)}(p) + (\Sigma(p)G^{(0)})^2 + \cdots),
\end{aligned}$$

reproducing term by term the expansion of $G(p)$. The infinite series can be summed analytically to give:

$$G(p) = \frac{G^{(0)}(p)}{1 - \Sigma(p)G^{(0)}(p)} = \frac{1}{G^{(0)}(p)^{-1} - \Sigma(p)} \tag{3.58}$$

Explicitly, the electron Green function reads:

$$G(\mathbf{k}, \omega) = \frac{1}{\omega - \epsilon_\mathbf{k} + i\eta - \Sigma(\mathbf{k}, \omega)} \tag{3.59}$$

where we have used the notation $p = (\mathbf{k}, \omega)$. Finding good approximations of the self-energy on a particular system is one of the main aims of quantum many-body theory.

Exercise I.8: Interacting Fermi gas at $T = 0$

Consider an interacting electron gas under the Coulomb interaction, $V(\mathbf{q})$.

$$H = \sum_{k,\sigma} \epsilon_\mathbf{k} c_{\mathbf{k}\sigma}^\dagger c_{\mathbf{k}\sigma} + \sum_{\mathbf{k},\mathbf{k}',\mathbf{q}} V(\mathbf{q}) c_{\mathbf{k}+\mathbf{q}\sigma}^\dagger c_{\mathbf{k}'-\mathbf{q}\sigma'}^\dagger c_{\mathbf{k}'\sigma'} c_{\mathbf{k}\sigma},$$

where the energies $\epsilon_\mathbf{k}$ are referred to the chemical potential.

(i)   Draw all Feynman diagrams for $G(\mathbf{k}, \omega)$ up to second order in $V(\mathbf{q})$.
(ii) Using FDPT in momentum evaluate explicitly the first order self-energy diagrams.
(iii) Obtain the spectral density $\rho_\sigma(\mathbf{k}, \omega)$ to first order in $V(\mathbf{q})$. Explain the effect of $V(\mathbf{q})$ comparing with the non-interacting spectral density.

Hint: the exchange diagram involves an integration in momentum which yields:

$$\int_0^{k_F} \frac{d\mathbf{k}_1}{(2\pi)^3} \frac{4\pi e^2}{|\mathbf{k} - \mathbf{k}_1|^2} = \frac{e^2 k_F}{\pi} \left(1 + \frac{k_F^2 - k^2}{2kk_F} \ln \left| \frac{k + k_F}{k - k_F} \right| \right)$$

3.10 Physical Interpretation of the Self-energy

As an electron propagates in an interacting Fermi gas, it produces particle-hole excitations due to Coulomb interaction. Among other effects, particle-hole excitations lead to a screening cloud so that the bare electron can be viewed as a quasiparticle with an effective mass. These kind of effects are encoded in the self-energy which contains a real and an imaginary part:

$$\Sigma(\mathbf{k}, \omega) = Re\Sigma(\mathbf{k}, \omega) + iIm\Sigma(\mathbf{k}, \omega) \tag{3.60}$$

where the real part, $\sigma^R(\mathbf{k}, \omega)$ has to do with shifts in the non-interacting energies $\epsilon_\mathbf{k}$ while $\sigma^R(\mathbf{k}, \omega)$ is related to the decay of the electron in particle-hole pairs.

The spectral function reads:

$$\rho(\mathbf{k}, \omega) = \mp\frac{1}{\pi}ImG^{R/A}(\mathbf{k}, \omega) = \mp\frac{1}{\pi}\frac{Im\Sigma^{R/A}(\mathbf{k}, \omega)}{(\omega - \epsilon_\mathbf{k} - Re\Sigma^{R/A}(\mathbf{k}, \omega))^2 + Im\Sigma^{R/A}(\mathbf{k}, \omega)^2} \tag{3.61}$$

where $Im\Sigma^R(\mathbf{k}, \omega) < 0$ so $\rho(\mathbf{k}, \omega) > 0$, as it should.

Note that since:

$$G^A(\mathbf{k}, \omega) = G^{R*}(\mathbf{k}, \omega) \tag{3.62}$$

we have the same relation between self-energies:

$$\Sigma^A(\mathbf{k}, \omega) = \Sigma^{R*}(\mathbf{k}, \omega). \tag{3.63}$$

We define the quasiparticle and inverse lifetime of the quasiparticles:

$$\tilde{\epsilon}_\mathbf{k} = \epsilon_\mathbf{k} + Re\Sigma(\mathbf{k}, \omega)|_{\omega=\tilde{\epsilon}_\mathbf{k}},$$
$$\Gamma(\mathbf{k}, \omega) = -Im\Sigma^R(\mathbf{k}, \omega)|_{\omega=\tilde{\epsilon}_\mathbf{k}} \tag{3.64}$$

In order to explore some general properties close to the Fermi energy we assume $\Sigma^R(\mathbf{k}, \omega)$ is small and expand:

$$\omega - \epsilon_\mathbf{k} - Re\Sigma(\mathbf{k}, \omega) \approx \omega - \tilde{\epsilon}_\mathbf{k} - \frac{\partial Re\Sigma(\omega)}{\partial \omega}|_{\omega=\tilde{\epsilon}_\mathbf{k}}(\omega - \tilde{\epsilon}_\mathbf{k}) \tag{3.65}$$

which can be re-expressed as:

$$\omega - \epsilon_\mathbf{k} - Re\Sigma(\mathbf{k}, \omega) \approx (\omega - \tilde{\epsilon}_\mathbf{k})(1 - \frac{\partial Re\Sigma(\omega)}{\partial \omega})|_{\omega=\tilde{\epsilon}_\mathbf{k}} = \frac{\omega - \tilde{\epsilon}_\mathbf{k}}{Z_\mathbf{k}} \tag{3.66}$$

where we have defined the quasiparticle weight:

$$Z_\mathbf{k} = \left(1 - \frac{\partial Re\Sigma(\omega)}{\partial \omega}\right)^{-1}|_{\omega=\tilde{\epsilon}_\mathbf{k}} \tag{3.67}$$

Hence, at low frequencies the Green functions of an interacting Fermi liquid reads:

$$G^{R/A}(\mathbf{k}, \omega) = \frac{Z_\mathbf{k}}{\omega - \tilde{\epsilon}_\mathbf{k} \pm i\tilde{\Gamma}_\mathbf{k}} \tag{3.68}$$

where:

$$\tilde{\epsilon}_{\mathbf{k}} = \epsilon_{\mathbf{k}} + Re\,\Sigma(\mathbf{k}, \omega)|_{\omega=\tilde{\epsilon}_{\mathbf{k}}}, \quad \tilde{\Gamma}(\mathbf{k}, \omega) = Z_{\mathbf{k}}\Gamma(\mathbf{k}, \tilde{\epsilon}_{\mathbf{k}}) \tag{3.69}$$

In a Fermi liquid the decay rate goes to zero at the Fermi energy:

$$Im\,\Sigma(k \to k_F, \omega \to 0) \to 0. \tag{3.70}$$

Luttinger (1961), in fact, demonstrated that at $T = 0$:

$$\tilde{\Gamma}_{\mathbf{k}}(\omega) \propto \omega^2. \tag{3.71}$$

to all orders in perturbation theory close to the Fermi energy, i.e., $\omega \to 0$.

Based on the above we can obtain a general expression for the spectral function of quasiparticles in a Fermi liquid at $k = k_F$:

$$\rho(\mathbf{k}_F, \omega) = Z_{k_F}\delta(\omega - \tilde{\epsilon}_{\mathbf{k}_F}) + \rho^{inc}(\mathbf{k}, \omega) \tag{3.72}$$

where $\tilde{\epsilon}_{\mathbf{k}_F} = 0$ and $\rho^{inc}(\mathbf{k}, \omega)$ denotes the incoherent contribution to the spectral density. The full Green function of an electron in a Fermi liquid is, in general:

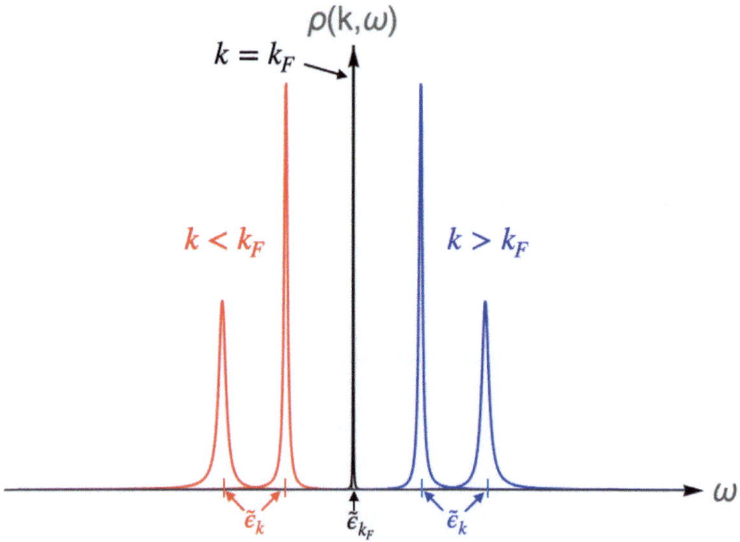

Fig. 3.2 Spectral function of quasiparticles in a Fermi liquid. As the energy of the quasiparticle, $\tilde{\epsilon}_k$, approaches the Fermi surface, $k < k_F$ (left) or $k > k_F$ (right) its decay rate, $\tilde{\Gamma}(\omega) \propto Z\omega \ll \tilde{\epsilon}_k$. At the Fermi surface, $k = k_F$, quasiparticles cannot decay into particle-hole pairs and so the spectral function is a Dirac delta function as in a non-interacting Fermi gas

$$G(\mathbf{k}, \omega \pm i\eta) = \frac{Z_{\mathbf{k}}}{\omega - \epsilon_{\mathbf{k}} \pm \Gamma_{\mathbf{k}}} + G^{inc}(\mathbf{k}, \omega) \tag{3.73}$$

where the first term is the Green function of the quasiparticles and the second the incoherent part. The spectral function of quasiparticles across the Fermi surface of a Fermi liquid is shown in Fig. 3.2.

Note that the quasiparticle weight may be more physically interpreted (since the self-energy is quite abstract entity) as:

$$Z_{\mathbf{k}} = |\langle QP, \mathbf{k}, \sigma | c_{\mathbf{k}\sigma}^{\dagger} | \Psi_G \rangle|^2, \tag{3.74}$$

which is the overlap between the added electron on the ground state of the Fermi liquid and the dressed electron or quasiparticle. Finally, Fourier transforming $G^{R/A}(\mathbf{k}, \omega) = G(\mathbf{k}, \omega \pm i\eta)$ we find the time dependent propagator:

$$G^{R/A}(\mathbf{k}, t) \approx \mp i e^{-i\tilde{\epsilon}_{\mathbf{k}}t} e^{\mp \tilde{\Gamma}_{\mathbf{k}}t} \theta(\pm t) Z_{\mathbf{k}}. \tag{3.75}$$

Since quasiparticles can decay due the Coulomb interaction they have a finite lifetime in the interval: $0 < t \leq \frac{1}{\tilde{\Gamma}_{\mathbf{k}}}$, in contrast to free fermions for which $\Gamma_{\mathbf{k}} = 0$.

Chapter 4
Finite Temperature Green Function Formalism

Developing a Green function formalism similar to the $T = 0$ FDPT is much more complicated. Kubo realized that the Boltzmann factor appearing in the evaluation of the Green function at finite-T can be treated as a time evolution operator by introducing imaginary times.

$$e^{-\beta H} = e^{-iH(-i\beta)t} = U(-i\beta, 0) \tag{4.1}$$

where $\beta = 1/k_B T$. Recall that the time evolution operator in Schrödingers picture is: $U(t, 0) = e^{-iHt}$. Hence, based on the above we can interpret the Boltzmann factor as evolution in imaginary times. This is a Wick rotation from real to imaginary time: $t \to -i\tau$ where τ is a real number in the interval $0 \le \tau \le \beta$ as we will discuss below. Lets now study how this substitution $t \to -i\tau$ is implemented in practical terms.

4.1 Schrödinger Representation

If we make the substitution $t \to -i\tau$ in Schrödingers equation:

$$H|\Psi_S\rangle = i\frac{\partial}{\partial t}|\Psi_S\rangle \tag{4.2}$$

we would find:

$$H|\Psi_S\rangle = \frac{\partial}{\partial\tau}|\Psi_S\rangle = -\frac{\partial}{\partial\tau}|\Psi_S\rangle. \tag{4.3}$$

so:

$$|\Psi_S(\tau)\rangle = e^{-H\tau}|\Psi_S(0)\rangle. \tag{4.4}$$

and the operator is time-independent.

J. Merino and A. L. Yeyati, *Many-Body Techniques in Condensed Matter Physics*, UNITEXT for Physics, https://doi.org/10.1007/978-3-031-55143-7_4

4.2 Heisenberg Representation

Operators in imaginary time under the transformation: $t \to -i\tau$ become:

$$O_H(\tau) = e^{H\tau} O_S e^{-H\tau} \tag{4.5}$$

Under derivation w.r.t τ we find the equation of motion:

$$\frac{\partial O_H(\tau)}{\partial \tau} = [H, O_H(\tau)] \tag{4.6}$$

and the time-independent state $|\Psi_H\rangle = |\Psi_S(0)\rangle$.

4.3 Imaginary Time Green Function

We introduce the imaginary time or Matsubara (Matsubara 1955) Green function appropriate at finite-T:

$$G(x, x') = G(\mathbf{x}, \tau, \mathbf{x}', \tau') \equiv -\langle T_\tau[\Psi(\mathbf{x}, \tau)\Psi^\dagger(\mathbf{x}', \tau')]\rangle = -\frac{1}{Z} Tr\{e^{-\beta H} T_\tau[\Psi(\mathbf{x}, \tau)\Psi^\dagger(\mathbf{x}', \tau')]\}$$
$$\tag{4.7}$$

where $T_\tau[]$ is the time ordering operator in imaginary times. We assume a uniform system in space and homogeneous in time, $H \neq H(t)$ as we did at $T = 0$. Hence, we can fix $\mathbf{x}' = 0, \tau' = 0$.

Periodicity/antiperiodicity:
Taking negative times: $-\beta < \tau < 0$:

$$G(\mathbf{k}, \tau) = \mp\langle\Psi_{\mathbf{k}\sigma}^\dagger(0)\Psi_{\mathbf{k}\sigma}(\tau)\rangle = \mp\frac{1}{Z} Tr\{e^{-\beta H}\Psi_{\mathbf{k}\sigma}^\dagger(0)e^{H\tau}\Psi_{\mathbf{k}\sigma}(0)e^{-H\tau}\}. \tag{4.8}$$

where the upper/lower sign refers to bosons/fermions. Using the cyclic property of the trace:

$$G(\mathbf{k}, \tau) = \mp\frac{1}{Z} Tr\{e^{H\tau}\Psi_{\mathbf{k}\sigma}(0)e^{-H\tau}e^{-\beta H}\Psi_{\mathbf{k}\sigma}^\dagger(0)\}$$

$$= \mp\frac{1}{Z} Tr\{e^{-\beta H}e^{H(\tau+\beta)}\Psi_{\mathbf{k}}(0)e^{-H(\tau+\beta)}\Psi_{\mathbf{k}\sigma}^\dagger(0)\}$$

$$= \mp\langle\Psi_{\mathbf{k}\sigma}(\tau + \beta)\Psi_{\mathbf{k}\sigma}^\dagger(0)\rangle = \pm G(\mathbf{k}, \tau + \beta). \tag{4.9}$$

Thus:

$$G(\mathbf{k}, \tau) = \pm G(\mathbf{k}, \tau + \beta), \tag{4.10}$$

for $\tau < 0$. This is a general property of finite-T Green function.

Example Imaginary time Green function of free fermions
Consider thee hamiltonian of the free Fermi gas:

$$H = \sum_{\mathbf{k}\sigma} \epsilon_{\mathbf{k}} c^{\dagger}_{\mathbf{k}\sigma} c_{\mathbf{k}\sigma}, \tag{4.11}$$

with energies referred to the chemical potential: $\epsilon_{\mathbf{k}} = \epsilon^0_{\mathbf{k}} - \mu$. From the Heisenberg equation of motion:

$$\frac{\partial c_{\mathbf{k}\sigma}(\tau)}{\partial \tau} = [H, c_{\mathbf{k}\sigma}(\tau)] = e^{H\tau}[H, c_{\mathbf{k}\sigma}]e^{-H\tau}. \tag{4.12}$$

Since:

$$[H, c_{\mathbf{k}\sigma}] = -\epsilon_{\mathbf{k}} c_{\mathbf{k}\sigma}, \tag{4.13}$$

we have:

$$\frac{\partial c_{\mathbf{k}\sigma}}{\partial \tau} = -\epsilon_{\mathbf{k}} c_{\mathbf{k}\sigma}(\tau), \tag{4.14}$$

which leads to the anihilation operator in imaginary time:

$$c_{\mathbf{k}\sigma}(\tau) = e^{-\epsilon_{\mathbf{k}}\tau} c_{\mathbf{k}\sigma}. \tag{4.15}$$

A similar derivation gives for the creation operator:

$$c^{\dagger}_{\mathbf{k}\sigma}(\tau) = e^{\epsilon_{\mathbf{k}}\tau} c^{\dagger}_{\mathbf{k}\sigma}. \tag{4.16}$$

Note that in contrast to real time Heisenberg operators now we have: $(c_{\mathbf{k}\sigma}(\tau))^{\dagger} \neq c^{\dagger}_{\mathbf{k}\sigma}$.

We now introduce $c_{\mathbf{k}\sigma}(\tau)$, $c^{\dagger}_{\mathbf{k}\sigma}(\tau)$ in the definition of the imaginary time Green function:

$$\begin{aligned}
G_{\sigma}(\mathbf{k}, \tau) &= -\langle T_{\tau}[c_{\mathbf{k}\sigma}(\tau), c^{\dagger}_{\mathbf{k}\sigma}(0)]\rangle = -\theta(\tau)\langle c_{\mathbf{k}\sigma}(\tau), c^{\dagger}_{\mathbf{k}\sigma}(0)\rangle + \theta(-\tau)\langle c^{\dagger}_{\mathbf{k}\sigma}(0) c_{\mathbf{k}\sigma}(\tau)\rangle \\
&= -\theta(\tau)e^{-\epsilon_{\mathbf{k}}\tau}(1 - \langle c^{\dagger}_{\mathbf{k}\sigma} c_{\mathbf{k}\sigma}\rangle) + \theta(-\tau)e^{-\epsilon_{\mathbf{k}}\tau}\langle c^{\dagger}_{\mathbf{k}\sigma} c_{\mathbf{k}\sigma}\rangle \\
&= -e^{-\epsilon_{\mathbf{k}}\tau}\{\theta(\tau)(1 - f(\epsilon_{\mathbf{k}})) - \theta(-\tau)f(\epsilon_{\mathbf{k}})\}.
\end{aligned} \tag{4.17}$$

So one can check that it satisfies the antiperiodicity property:

$$G_{\sigma}(\mathbf{k}, \tau + \beta) = -G_{\sigma}(\mathbf{k}, \tau), \tag{4.18}$$

for $-\beta < \tau < 0$ (Fig. 4.1).
    It is instructive to plot this function.

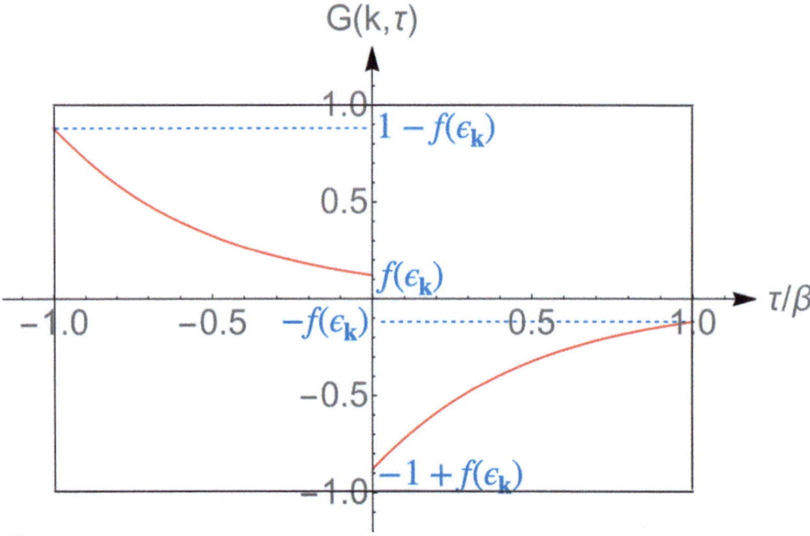

Fig. 4.1 Imaginary time Green function of a free fermion. $G(\mathbf{k}, \tau)$ given in Eq. (4.17) is plotted for an electron with energy, $\epsilon_{\mathbf{k}} = 0.2$, and $\beta = 10$. Note the antiperiodicity of $G(\tau)$ in the imaginary time interval, $-\beta \leq \tau \leq \beta$, and the relation of $G(\mathbf{k}, \tau)$ with $f(\epsilon_{\mathbf{k}})$

4.3.1 Matsubara Representation of the Imaginary Time Green Function

Due to the periodicity/antiperiodicity of the imaginary time Green function we can expand it in a Fourier series with discrete frequencies:

$$G(\mathbf{k}, \tau) = \frac{1}{\beta} \sum_n G(\mathbf{k}, i\alpha_n) e^{-i\alpha_n \tau}, \tag{4.19}$$

where:

$$\alpha_n = \nu_n = \frac{\pi 2n}{\beta} \text{ for bosons}$$

$$\alpha_n = \omega_n = \frac{\pi(2n + 1)}{\beta} \text{ for fermions} \tag{4.20}$$

with the inverse Fourier transform:

$$G(\mathbf{k}, i\alpha_n) = \frac{1}{2} \int_{-\beta}^{\beta} d\tau e^{i\alpha_n \tau} G(\mathbf{k}, \tau) d\tau = \int_0^{\beta} e^{i\alpha_n \tau} G(\mathbf{k}, \tau) d\tau. \tag{4.21}$$

Example Matsubara representation of free fermion Green function.

We consider again the free fermion gas described by model (4.11). Using the imaginary time Green function (4.17) in the expression for the inverse Fourier transform we find:

$$G(\mathbf{k}, i\omega_n) = -\int_0^\beta d\tau e^{(i\omega_n - \epsilon_\mathbf{k})\tau}(1 - f(\epsilon_\mathbf{k})) = \frac{1}{i\omega_n - \epsilon_\mathbf{k}}. \tag{4.22}$$

This is the Matsubara representation of the free fermion imaginary time Green function. It has a very similar form to the $T = 0$ causal Green function found earlier if we replace $i\omega_n \to \omega$ and we also suppress $i\eta_\mathbf{k}$. However, we will see below that there is a direct general relation between the Matsubara and the finite temperature retarded Green function which is the experimentally measured quantity.

4.4 Spectral Decomposition of Matsubara Green Function

We now obtain the spectral decomposition of the Matsubara Greens function similarly to the Lehmann representation we found for the $T = 0$ Green function. This representation is general for any system of interacting particles assuming that the orthonormal set of eigenvalues, $\{|n\rangle\}$, and eigenstates, E_n, of the system with any particle number is known.

Using the inverse Fourier transform:

$$\begin{aligned}
G_\sigma(\mathbf{k}, i\alpha_n) &= \mp \int_0^\beta e^{i\alpha_n\tau} \langle c_\mathbf{k}(\tau)c_{\mathbf{k}\sigma}^\dagger(0)\rangle \\
&= -\frac{1}{Z}\int_0^\beta d\tau e^{i\omega_n\tau} \sum_{n,m} e^{-\beta E_n} \langle n|e^{H\tau}c_{\mathbf{k}\sigma}e^{H\tau}|m\rangle\langle m|c_\sigma^\dagger(\mathbf{k})|n\rangle \\
&= -\frac{1}{Z}\sum_{n,m} e^{-\beta E_n}|\langle n|c_{\mathbf{k}\sigma}|m\rangle|^2 \int_0^\beta d\tau e^{(i\omega_n + E_N - E_m)\tau} \\
&= -\frac{1}{Z}\sum_{n,m} \frac{|\langle n|c_{\mathbf{k}\sigma}|m\rangle|^2}{i\alpha_n + E_n - E_m}(e^{-\beta E_m} \pm e^{-\beta E_n}) \tag{4.23}
\end{aligned}$$

where the upper (lower) signs with their corresponding Matsubara frequencies $\alpha_n = \omega_n(\nu_n)$ refer to fermions (bosons).

On the other hand we can also obtain a spectral decomposition of the real time finite-T retarded/advanced Green functions defined as:

$$G_\sigma^{R/A}(\mathbf{k}, t) \equiv \mp i\langle\{c_{\mathbf{k}\sigma}(t), c_{\mathbf{k}\sigma}^\dagger(0)\}\rangle\theta(\pm t), \tag{4.24}$$

and for bosons:

$$G_\sigma^{R/A}(\mathbf{k}, t) \equiv \mp i\langle[c_{\mathbf{k}\sigma}(t), c_{\mathbf{k}\sigma}^\dagger(0)]\rangle\theta(\pm t). \tag{4.25}$$

Applying the same procedure for the real time retarded Green function as for the Matsubara Greens function gives:

$$G^R(\mathbf{k}, \omega) = \frac{1}{Z} \sum_{n,m} \frac{|\langle n|c_{\mathbf{k}\sigma}|m\rangle|^2}{\omega + i\eta - (E_m - E_n)} (e^{-\beta E_n} \pm e^{-\beta E_m}). \qquad (4.26)$$

Comparing with the previous spectral decomposition of the Matsubara Green function we conclude that:

$$G^{R/A}(\mathbf{k}, \omega) = G(\mathbf{k}, \omega \pm i\eta). \qquad (4.27)$$

This is a crucial relation since observable properties of the system given by $G^R(\omega)$ can be obtained by evaluating the Matsubara Green function at the analytically continued frequencies: $i\omega_n \to \omega + i\eta$. So if we now how to compute the Matsubara Green function we can obtain observable properties such as ARPES just by analytical continuation of $G(i\omega_n)$ to the real axis.

The finite-T spectral function can easily be obtained:

$$\rho_\sigma(\mathbf{k}, \omega) = -\frac{1}{\pi} Im G^R_\sigma(\mathbf{k}, \omega) = -\frac{1}{\pi} Im G_\sigma(\mathbf{k}, \omega + i\eta)$$

$$= \frac{1}{Z} \sum_{n,m} |\langle n|c_{\mathbf{k}\sigma}|m\rangle|^2 e^{-\beta E_n} (1 \pm e^{-\beta\omega})\delta(\omega - (E_m - E_n)). \quad (4.28)$$

One can show that $\rho_\sigma(\mathbf{k}, \omega)$ satisfies the sum rule:

$$\int_{-\infty}^{\infty} \rho_\sigma(\mathbf{k}, \omega) = 1, \qquad (4.29)$$

as in the $T = 0$ case. The Matsubara, retarded and advanced Greens have a spectral decomposition in terms of the spectral function:

$$G_\sigma(\mathbf{k}, i\omega_n) = \int_{-\infty}^{\infty} d\omega' \frac{\rho_\sigma(\mathbf{k}, \omega')}{i\omega_n - \omega'}$$

$$G^{R/A}_\sigma(\mathbf{k}, \omega) = \int_{-\infty}^{\infty} d\omega' \frac{\rho_\sigma(\mathbf{k}, \omega')}{\omega \pm i\eta - \omega'} = G_\sigma(\mathbf{k}, \omega \pm i\eta). \qquad (4.30)$$

4.5 Perturbation Theory at Finite Temperature

We are interested in treating the interaction part, V, of the hamiltonian:

$$H = H_0 + V \qquad (4.31)$$

perturbatively. For this purpose we introduce the interaction representation as in the $T = 0$ formalism.

4.5.1 Interaction Representation

Introducing the Wick rotation $t \rightarrow -i\tau$, the states in the interaction representation in imaginary time reads:

$$|\Psi_I(\tau)\rangle = e^{H_0\tau}|\Psi_s(\tau)\rangle = e^{(H_0-H)\tau}|\Psi_H\rangle, \tag{4.32}$$

in terms of the Schrödinger and Heisenberg states. The operators in the interaction representation in terms of the Heisenberg read:

$$O_H(\tau) = U(\tau, 0)^{-1} O_I(\tau) U(\tau, 0) \tag{4.33}$$

where the time evolution operator in the interaction representation in imaginary times connecting states at different times:

$$|\Psi_I(\tau)\rangle = U(\tau, \tau')|\Psi_I(\tau')\rangle \tag{4.34}$$

has the explicit form:

$$U(\tau, \tau') = e^{H_0\tau} e^{-H(\tau-\tau')} e^{-H_0\tau'} \tag{4.35}$$

and satisfies the equation of motion:

$$-\frac{\partial}{\partial\tau} U(\tau, \tau') = V_I(\tau) U(\tau, \tau'). \tag{4.36}$$

The time-evolution operator can be expanded in a similar way as in the $T = 0$ formalism:

$$U(\tau, \tau') = T exp\{-\int_{\tau'}^{\tau} V_I(\tau_1) d\tau_1\} \tag{4.37}$$

which is the starting point of the perturbation series expansion of the Green function.

4.5.2 Perturbation Theory and Wick's Theorem

We consider the Green function for $\tau > \tau'$:

$$G(\mathbf{k}, \tau - \tau') = -\langle c_{\mathbf{k}}(\tau) c_{\mathbf{k}}(\tau')\rangle = -\frac{1}{Z} Tr\{e^{-\beta H} c_{\mathbf{k}}(\tau) c_{\mathbf{k}}^{\dagger}(\tau')\}. \tag{4.38}$$

Which re-expressed in the interaction representation reads:

$$G(\mathbf{k}, \tau - \tau') = -\frac{1}{Z}\{e^{-\beta H_0}e^{\beta H_0}e^{-\beta H}U(0, \tau)c_\mathbf{k}(\tau)U(\tau, 0)U(0, \tau')c_\mathbf{k}^\dagger(\tau')U(\tau', 0)\}$$

(4.39)

and the partition function can be rewritten as:

$$Z = Tr\{e^{-\beta H}\} = Tr\{e^{-\beta H_0}e^{\beta H_0}e^{-\beta H}\}$$

(4.40)

or, equivalently:

$$Z = Tr\{e^{-\beta H}\} = Tr\{e^{-\beta H_0}U(\beta, 0)\}$$

(4.41)

For $\tau > \tau'$ we find:

$$G(\mathbf{k}, \tau - \tau') = -\frac{1}{Z}Tr\{e^{-\beta H_0}U(\beta, 0)U(0, \tau)c_\mathbf{k}(\tau)U(\tau, \tau')c_\mathbf{k}^\dagger(\tau')U(\tau', 0)\}$$

$$= -\frac{1}{Z}Tr\{e^{-\beta H_0}T_\tau[U(\beta, 0)c_\mathbf{k}(\tau)c_\mathbf{k}^\dagger(\tau')\}$$

(4.42)

where $U(\beta, 0)$ is the S-matrix. Performing the expansion on the S-matrix we have:

$$G(\mathbf{k}, \tau - \tau') = -\frac{1}{Z}Tr\{e^{-\beta H_0}\sum_{n=0}^{\infty}\frac{(-1)^n}{n!}\int_0^\beta d\tau_1...\int_0^\beta d\tau_n T_\tau[V(\tau_1)...V(\tau_n)c_\mathbf{k}(\tau)c_\mathbf{k}^\dagger(\tau')]\}$$

(4.43)

And the expansion of the partition function:

$$Z = Tr\{e^{-\beta H_0}\sum_{n=0}^{\infty}\frac{(-1)^n}{n!}\int_0^\infty d\tau_1\int_0^\infty d\tau_n T_\tau[V(\tau_1)...V(\tau_n)]\}$$

(4.44)

A diagrammatic analysis similar to the $T = 0$ case can be performed. Since the numerator in the Green function of Eq. (4.43) can be factorized as:

$$Tr\{e^{-\beta H_0}T_\tau[Sc_\mathbf{k}(\tau)c_\mathbf{k}^\dagger(\tau')]\} = Tr\{e^{-\beta H_0}T_\tau[Sc_\mathbf{k}(\tau)c_\mathbf{k}^\dagger(\tau')]\}_{conn.}Tr\{e^{-\beta H_0}S\}.$$

(4.45)

The $Tr\{e^{-\beta H_0}S\}$ cancels Z in the denominator of Eq. (4.43) and so:

$$G(\mathbf{k}, \tau - \tau') = -Tr\{e^{-\beta H_0}T_\tau[Sc_\mathbf{k}(\tau)c_\mathbf{k}^\dagger(\tau')]\}_{conn.}$$

(4.46)

4.6 Feynman Rules for the Coulomb Interaction at Finite-T

At $T \neq 0$ the Feynman rules in momentum space for the Coulomb interaction read:

1. Draw all topologically distinct connected diagrams with n interaction lines and $2n + 1$ directed particle lines.

2. Associate momenta and Matsubara frequencies with the lines of the diagrams in such a way that external lines carry the external momentum and frequency, say, $p = (\mathbf{p}, \omega_n)$ while the momenta and frequencies of internal lines satisfy the conservation laws, $\sum_i \mathbf{p}_i = 0$, $\sum_{n_i} \omega_{n_i} = 0$, where frequencies of Bose lines are always even, $\nu_n = \frac{2n\pi}{\beta}$ and of Fermi lines are odd, $\omega_n = \frac{(2n+1)\pi}{\beta}$.

3. With each solid internal line of momentum $\mathbf{p}_i$ and frequency ω_{n_i} associate the non-interacting propagator:

$$\frac{1}{i\omega_{n_i} + \mu - \epsilon^0_{\mathbf{p}_i}} \tag{4.47}$$

and with each wavy line associate the quantity: $V(q) = V(\mathbf{q}, \nu_n)$.

4. Integrate over all internal momenta and frequencies.

5. With both external lines associate:

$$\frac{\delta_{\alpha\beta}}{i\omega_n + \mu - \epsilon^0_{\mathbf{p}}} \tag{4.48}$$

6. Multiply the expression by:

$$(-1)^n \frac{\beta^{-n}}{(2\pi)^{3n}} (-1)^F (2S + 1)^F \tag{4.49}$$

7. A solid line with identical time arguments should have a convergence factor $e^{i\omega_n 0^+}$ inserted.

4.7 Feynman Diagrams at Finite Temperature

Let's consider, as an example, the first order corrections, $n = 1$, to $G^{(0)}(\mathbf{p}, i\omega_n)$. Diagrammatically:

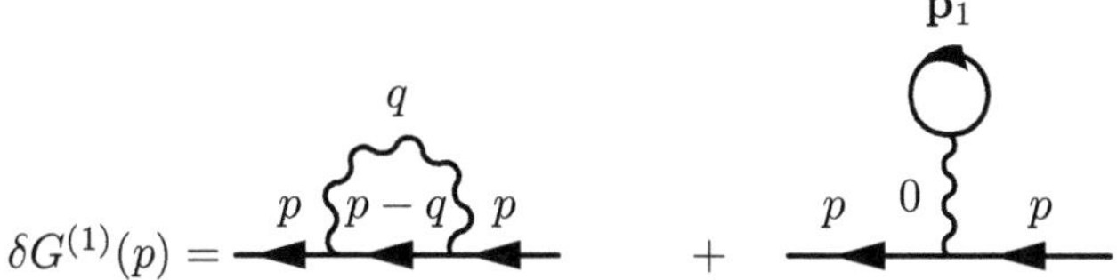

$$\delta G^{(1)}(p) = \quad + \quad$$

The Hartree-Fock contributions would give, applying the Feynman rules:

$$\delta G^{(1)}(\mathbf{p}, i\omega_n) =$$

$$= \frac{1}{\beta}(-1)(-1)V(0)(2S+1)\frac{\delta_{\alpha\beta}}{(i\omega_n + \mu - \epsilon_{\mathbf{p}}^0)^2}\sum_{n_1}\int \frac{d\mathbf{k}_1}{(2\pi)^3}\frac{e^{i\omega_{n_1}0^+}}{i\omega_{n_1} + \mu - \epsilon_{\mathbf{k}_1}^0}$$

$$-\frac{1}{\beta}\frac{\delta_{\alpha\beta}}{(i\omega_n + \mu - \epsilon_{\mathbf{p}}^0)^2}\sum_{n_1}\int \frac{d\mathbf{k}_1}{(2\pi)^3}e^{i\omega_{n_1}0^+}V(\mathbf{k}-\mathbf{k}_1)\frac{1}{i\omega_{n_1} + \mu - \epsilon_{\mathbf{k}_1}^0}.$$

4.7.1 Self-consistent Hartree-Fock Approach

The Dyson equation at finite-T is a straightforward extension of the $T = 0$ expression to imaginary frequencies:

$$G(\mathbf{k}, i\omega_n) = G^{(0)}(\mathbf{k}, i\omega_n) + G^{(0)}(\mathbf{k}, i\omega_n)\Sigma(\mathbf{k}, i\omega_n)G(\mathbf{k}, i\omega_n), \qquad (4.50)$$

which allows to express the the Greens function in terms of the exact irreducible self-energy, $\Sigma(\mathbf{k}, i\omega_n)$. Often we are interested in finding a sufficiently accurate approximation for the self-energy to describe our system.

In many cases, first order Hartree-Fock terms are a poor approximation for the Green function. Nevertheless, it can be improved by replacing the bare Green function, $G^{(0)}$, by the full propagator, G. In this case the self-energy is a functional of G:

$$\Sigma_{HF}[G] = \Sigma_H[G] + \Sigma_X[G] \qquad (4.51)$$

This leads to a set of self-consistent equations since G is coupled to $\Sigma_{HF}[G]$. We first give a guess for G and obtain a self-energy which is used to obtain a new G until self-consistency is achieved. This is the self-consistent Hartree-Fock approach. It resums a whole class of diagrams but only of the Hartree-Fock type.

Phase transitions and broken symmetry solutions that cannot be reached through perturbation theory around the normal state may still be found through an extension of the Green functions to a generalized matrix propagators including anomalous (beyond "normal") propagators describing long range (ferromagnetism, superconductivity,...) broken symmetry solutions. This allows to deal with the cases in which:

$$\langle \Phi_G | \Psi_G \rangle = 0 \qquad (4.52)$$

i.e. the ground state of the interacting system has a different symmetry from the ground state of the non-interacting system. This is not considered in the adiabatic hypothesis underlying the standard perturbation theory for which $\langle \Phi_G | \Psi_G \rangle \neq 0$.

Exercise 1.9 Quasiparticles in a Fermi liquid

The elementary excitations in the Landau theory of the Fermi liquid are quasi-particles characterized by the same quantum numbers, $\mathbf{k}, \sigma$, as those of the free electron gas (in spite of the large Coulomb repulsion between electrons) but with renormalized energies. Quasiparticles are well defined since their lifetime, τ_k, becomes infinite approaching the Fermi surface: $\Gamma_{k \to k_F}(\omega \to 0) = 1/\tau_{k \to k_F} \to 0$, where $\Gamma_k(\omega) = -2Im\Sigma^R(\mathbf{k}, \omega)$, is the relaxation rate with $\Sigma^R(\mathbf{k}, \omega)$, the retarded self-energy.

(i) Using Feynman rules at finite T write down an expression for the self-energy, $\Sigma(\mathbf{k}, i\omega)$, with Feynman diagram:

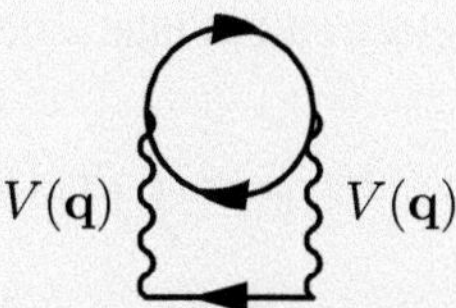

     neglecting the momentum dependence of $V(\mathbf{q}) = U$.

(ii) Obtain the imaginary part of $\Sigma^R(\mathbf{k}, \omega)$ and show that $\Gamma_{k_F}(\omega) = -2Im\Sigma^R(\mathbf{k}_F, \omega) \propto \omega^2$.

(iii) Show that the Green's function of an electron at the Fermi surface, $|\mathbf{k}| = k_F$ at $T = 0$ reads:

$$G_\sigma^R(\mathbf{k}, \omega) = \frac{Z_\mathbf{k}}{\omega + i\eta - \epsilon_\mathbf{k}^*},$$

giving an explicit expression for the quasiparticle weight: $Z_\mathbf{k}$ and the renormalized quasiparticle energy, $\epsilon_\mathbf{k}^*$, in terms of the retarded self-energy $\Sigma^R(\mathbf{k}, \omega)$.

(iv) Obtain the momentum distribution of the Fermi liquid:

$$n_{\mathbf{k}\sigma} = G_\sigma(\mathbf{k}, 0^-),$$

where $G_\sigma(\mathbf{k}, 0^-)$ is the Matsubara Green functions. Show that at $T = 0$, $n_{\mathbf{k}\sigma}$ has a discontinuous step at the Fermi momentum, k_F, i.e. $\epsilon_{k_F} = \epsilon_F$. and that the step height is equal to Z_k. Discuss your results.

4.8 Diagrammatic Approach to the Electron-Phonon Interaction

We have mainly discussed the Coulomb interaction. We now want to analyze the electron-phonon interaction perturbatively within the Matsubara Green function formalism. As a first step we derive the electron-phonon interaction in the second quantization. Then we will derive the diagrammatic approach for the electron-phonon interaction.

4.8.1 Electron-Phonon Interaction in Second Quantization

The starting point is the electron-ion interaction in a solid:

$$H_{e-ph} = \int d\mathbf{r}\rho(\mathbf{r})V_{ion}(\mathbf{r}), \tag{4.53}$$

where the potential of the ions reads:

$$V_{ion}(\mathbf{r}) = \sum_j \phi(\mathbf{r} - \mathbf{r}_j). \tag{4.54}$$

where $\mathbf{r}_j$ denotes the instantaneous position of the j-ion. The second quantized form of $H_{e-phonon}$ can be obtained expressing the electron density, $\rho(\mathbf{r})$, in terms of the field operators, $\Psi_\sigma(\mathbf{r})$:

$$\rho(\mathbf{r}) = -en(\mathbf{r}) = -e\sum_\sigma \Psi_\sigma^\dagger(\mathbf{r})\Psi_\sigma(\mathbf{r}). \tag{4.55}$$

where we take $e > 0$.

We want to analyze ion vibrations within the harmonic approximation. We assume small deviations from the equilibrium position of the ions:

$$\mathbf{r}_j = \mathbf{R_j} + \mathbf{u}_j, \tag{4.56}$$

with $|\mathbf{u}_j| << |\mathbf{R}_i - \mathbf{R}_j|$. By inserting this in V_{ion} we obtain the harmonic approximation to the electron-phonon interaction:

$$H_{e-phonon} = \int dr\rho(r) \sum_{j=1}^{N}(V_{ion}(r - R_j) + u_j\frac{\partial}{\partial x}V(r - x)|_{x=R_j} + O(u_j^2)) \tag{4.57}$$

where for simplicity we have assumed a 1D system. The Fourier transform of the spatial dependent quantities is given by:

$$\rho(r) = -\frac{e}{V} \sum_\sigma \sum_{k,k'} e^{-i(k-k')r} c_{k\sigma}^\dagger c_{k'\sigma}$$

$$u_j = \frac{1}{\sqrt{N}} \sum_k u_k e^{ikR_j}$$

$$V_{ion}(r) = \frac{1}{V} \sum_q V_q e^{iqr} \tag{4.58}$$

with $V = L^d$ ($d = 1$ and L the length of the system). Introducing these transformations in $H_{e-phonon}$ we have:

$$
\begin{aligned}
H_{e-ph} = {}& -\frac{e}{V} \int dr \sum_{\sigma,k,k'} e^{-i(k-k')r} c_{k\sigma}^\dagger c_{k'\sigma} \sum_j \left(\frac{1}{V} \sum_q V_q e^{iq(r-R_j^0)} \right. \\
& \left. - \frac{i}{\sqrt{N}V} \sum_{k''} u_{k''} e^{ik''R_j} \sum_q q V_q e^{iq(r-R_j)} \right) \\
= {}& -\frac{e}{V^2} \sum_{\sigma,k,k',q} V_q c_{k\sigma}^\dagger c_{k'\sigma} \int dr \sum_j e^{-i(k-k'-q)r} e^{-iqR_j} \\
& + i\frac{e}{V^2\sqrt{N}} \sum_{\sigma,k,k',k'',q} c_{k\sigma}^\dagger c_{k'\sigma} u_{k''} V_q q \int dr \sum_j e^{-i(k-k'-q)r} e^{i(k''-q)R_j}
\end{aligned}
$$

Performing the integration in r and the sum in j:

$$
\begin{aligned}
H_{e-ph} = {}& -\frac{e}{V^2} \sum_{\sigma,k,k',q} V_q c_{k\sigma}^\dagger c_{k'\sigma} V\delta(k - k' - q) N\delta(q) \\
& + i\frac{e}{\sqrt{N}V^2} \sum_{\sigma,k,k',k'',q} c_{k\sigma}^\dagger c_{k'\sigma} u_{k''} V_q q V\delta(k - k' - q) N\delta(k'' - q) \\
= {}& -eNV_0 \frac{1}{V} \sum_{k'\sigma} c_{k'\sigma}^\dagger c_{k'\sigma} \\
& + ie\frac{\sqrt{N}}{V} \sum_{\sigma,k',q} c_{k'+q\sigma}^\dagger c_{k'\sigma} u_q V_q q. \tag{4.59}
\end{aligned}
$$

The first term is just proportional to the density of the electron gas which we assume uniform so it is a constant term which we can neglect. Using the quantized fields associated with the vibrations of the ions:

$$u_q = \sqrt{\frac{\hbar}{2m\omega_q}} (b_{-q}^\dagger + b_q) \tag{4.60}$$

in the expression for H_{e-ph} we have:

$$H_{e-ph} = ie\frac{\sqrt{N}}{V}\sum_q\sum_{k'\sigma}c^{\dagger}_{k'+q\sigma}c_{k'\sigma}\sqrt{\frac{\hbar}{2m\omega_q}}(b^{\dagger}_{-q}+b_q)V_q q$$

$$= \sum_q g_q \sum_{k'\sigma}c^{\dagger}_{k'+q\sigma}c_{k'\sigma}(b^{\dagger}_{-q}+b_q) \qquad (4.61)$$

where the electron-phonon coupling reads:

$$g_q = ie\frac{\sqrt{N}}{V}\sqrt{\frac{\hbar}{2m\omega_q}}V_q q. \qquad (4.62)$$

In summary, the electron-phonon interaction in second quantization reads:

$$H_{e-ph} = \sum_q g_q \sum_{k'\sigma}c^{\dagger}_{k'+q\sigma}c_{k'\sigma}(b^{\dagger}_{-q}+b_q) \qquad (4.63)$$

Exercise 1.10 Finite temperature Green function for phonons
Consider the hamiltonian of a simple harmonic oscillator:

$$H_0 = \omega(b^{\dagger}b + \frac{1}{2}).$$

The imaginary time Green function of a single quantum harmonic oscillator is defined as:

$$D(\tau) = -\langle T_\tau x(\tau)x(0)\rangle,$$

where the quantum harmonic oscillator displacement, x, is expressed in terms of bosonic operators as:

$$x = x(0) = \sqrt{\frac{\hbar}{2m\omega}}(b + b^{\dagger}).$$

Obtain the expression for the quantum harmonic oscillator propagator in Matsubara frequencies, $D(i\omega_n)$.

4.8.2 *Feynman Diagrams*

From the above analysis we introduce the Fröhlich hamiltonian for a three dimensional Fermi gas in the presence of a vibrating lattice. This is our starting point for the diagrammatic approach of the electron-phonon problem. The Fröhlich hamiltonian(Frölich 1954) reads:

$$H_e = \sum_{\mathbf{k}\sigma} \epsilon_{\mathbf{k}} c^{\dagger}_{\mathbf{k}\sigma} c_{\mathbf{k}\sigma}$$

$$H_{phonon} = \sum_{\mathbf{k}\sigma} \omega_{\mathbf{k}\lambda} b^{\dagger}_{\mathbf{k}\lambda} b_{\mathbf{k}\lambda}$$

$$H_{e-ph} = \sum_{\mathbf{q}\lambda} g_{\mathbf{q}\lambda} c^{\dagger}_{\mathbf{k}+\mathbf{q}\sigma} c_{\mathbf{k}\sigma} (b_{\mathbf{q}\lambda} + b^{\dagger}_{-\mathbf{q}\lambda}) \tag{4.64}$$

where $\lambda = 1, 2, 3$ describes the three phonon modes of a 3D solid (one longitudinal and two transversal). In the discussion we will only keep the longitudinal mode. The two processes in the electron-phonon interaction are phonon absorption and emission.

To order $2n$ in the electron-phonon coupling, the contribution to the Green function can be obtained by the following rules:

1. Associate a solid line of momentum k to the non-interacting Green function:

$$G^{(0)}(\mathbf{k}, i\omega_n) = \frac{1}{i\omega_n - \epsilon_{\mathbf{k}}} \tag{4.65}$$

2. We associate to each ondulated line the phonon Green function:

$$D(\mathbf{k}, i\omega_n) = \frac{2\omega_{\mathbf{q}}}{(i\nu_n)^2 - \omega_{\mathbf{q}}^2} \tag{4.66}$$

3. Associate with each electron-phonon vertex scattering an electron from state $\mathbf{k}$ to state $\mathbf{k} + \mathbf{q}$ an electron-phonon coupling $i g_{\mathbf{q}}$.

By combining 2. and 3. we find there is an effective interaction induced by phonon exchange:

$$- V_{eff}(\mathbf{q}, i\nu_n) = (i g_{\mathbf{q}})^2 \frac{2\omega_{\mathbf{q}}}{(i\nu_n)^2 - \omega_q^2} \tag{4.67}$$

which is similar to the electron-electron Coulomb interaction but is frequency dependent.

4. Multiply the contribution of the diagram by:

$$\frac{1}{\beta^n} \frac{1}{(2\pi)^{3n}} (-1)^F (2S + 1)^F \tag{4.68}$$

5. Sum over all internal frequencies and integrate over all internal momenta.
6. Insert a $e^{i\omega_{n_i} 0^+}$ factor for each closed loop.

As an example to illustrate the Feynman rules let's evaluate the correction to the phonon propagator due to the excitation of particle-hole pairs. This process is very important as is related with the polarization of the Fermi gas due to phonons. It can be represented by the bubble diagram:

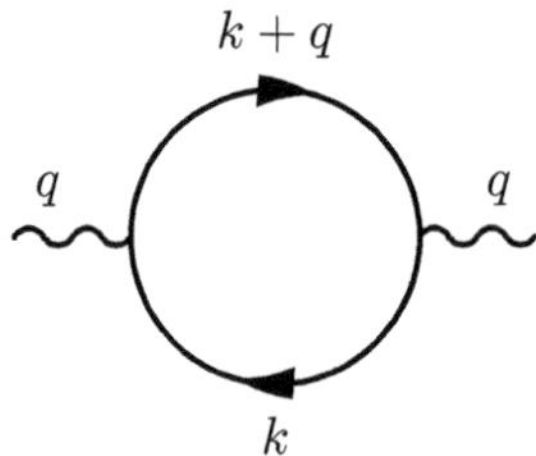

This diagram corresponds to $2n = 2$ order in the electron-phonon coupling. The corresponding correction to $D(\mathbf{q}, i\omega_n)$ reads:

$$(-1)(2S+1)(\frac{2\omega_{\mathbf{q}}}{(i\nu_n)^2 - \omega_{\mathbf{q}}^2})^2(ig_{\mathbf{q}})^2\frac{1}{\beta}\sum_m\int\frac{d\mathbf{k}}{(2\pi)^3}G^{(0)}(\mathbf{k}, i\omega_m)G^{(0)}(\mathbf{k}+\mathbf{q}, i\omega_m + i\nu_n)$$

$$= (2S+1)(\frac{2\omega_{\mathbf{q}}}{(i\nu_n)^2 - \omega_{\mathbf{q}}^2})^2\frac{1}{\beta}\sum_m\int\frac{d\mathbf{k}}{(2\pi)^3}\frac{1}{i\omega_m + \mu - \epsilon_{\mathbf{k}}}\frac{1}{i\omega_m + i\nu_n + \mu - \epsilon_{\mathbf{k}+\mathbf{q}}} \qquad (4.69)$$

Note that in order to calculate the contribution from this diagram the sum in Matsubara frequencies must still be evaluated.

4.9   Evaluation of Matsubara Sums

When using Matsubara Green functions one is generally faced with the task of performing sums over the discrete Matsubara frequencies. This can be done by using a contour integration technique in the complex plane, based on the analytical properties of the Fermi, $f(z)$, and/or Bose functions, $b(z)$. While $f(z)$ has simple poles at each $z = i\omega_n$, $b(z)$ at $z = i\nu_n$.

Lets consider the Fermi distribution with complex argument z:

$$f(z) = \frac{1}{e^{\beta z} + 1}, \qquad (4.70)$$

which has simple poles when: $e^{\beta z} + 1 = 0$, i. e., for $z = i\omega_n$. The residues at these poles can be easily found identifying $f(z)$ with $f(z) = \frac{g(z)}{h(z)}$ where $g(z) = 1$ and $h(z) = e^{\beta z} + 1$. Expanding $h(z)$ around the poles:

$$h(z) = e^{\beta z} + 1 \approx (z - i\omega_n)(-\beta - \beta^2\frac{(z - i\omega_n)}{2!} + \ldots) = (z - i\omega_n)r(z) \quad (4.71)$$

where $r(z)$ is analytical at $r(z = i\omega_n) = -\beta$.

The residues of $f(z)$ at $z = i\omega_n$ read:

$$\text{Res}(f, i\omega_n) = \lim_{z \to i\omega_n}\frac{g(z)}{h(z)}(z - i\omega_n) = \frac{g(z)}{r(z)}|_{z=i\omega_n} = -\frac{1}{\beta}, \qquad (4.72)$$

and are independent of n. From the residues of $f(z)$ at $z = i\omega_n$ we have that:

$$\frac{1}{\beta} \sum_n F(i\omega_n) = -\int_C \frac{dz}{2\pi i} F(z) f(z) = I \tag{4.73}$$

where C is the closed contour in Fig. 4.2. This relation allows the evaluation of Matsubara sums of a generic function $F(z)$ using contour integration.

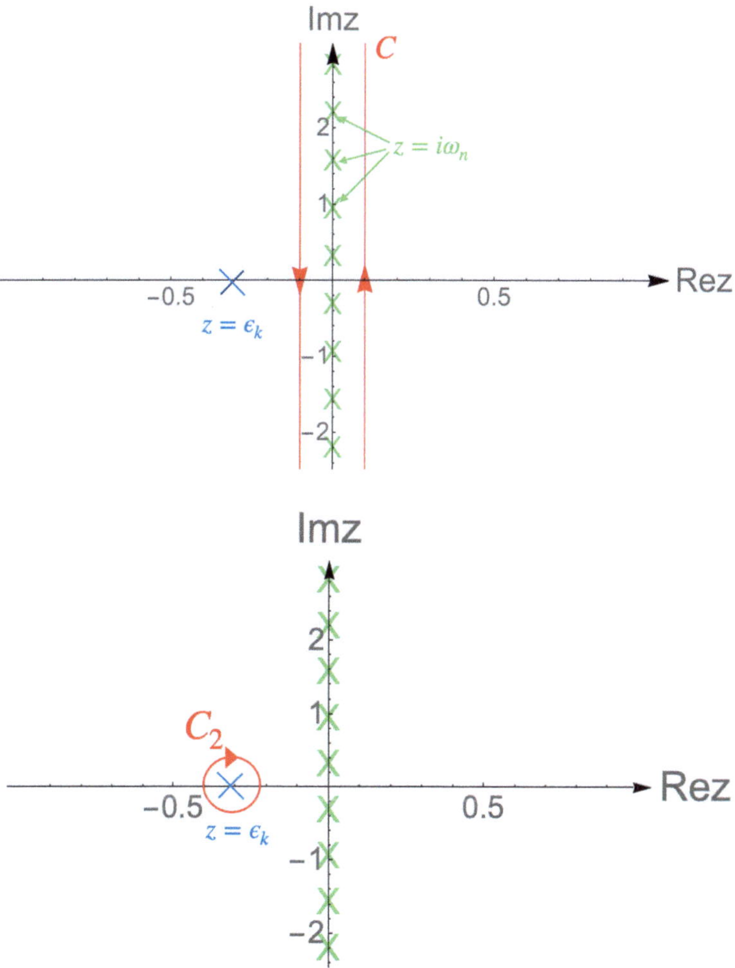

Fig. 4.2 Contours for evaluating a Matsubara sum through complex plane integration. The integration along the C contour surrounding only the imaginary Matsubara frequencies is converted into the C_2 contour around the simple pole in $F(z)$. This allows the evaluation of the occupation number at finite temperature

To evaluate this integral we may distort the contour, C, into an equivalent one, C', but with ∞ radius which encircles the simple poles and branch cuts of $F(z)$. Thus the contour can be expressed as: $C' = C_1 + C_2 + C_3$ where C_1 corresponds to the infinite radius section, C_2 to the section encircling the simple poles and C_3 the section along the branch cut. Thus, the integral can also be expressed as:

$$I = -\int_{C'} \frac{dz}{2\pi i} F(z) f(z) = -\int_{C_1} \frac{dz}{2\pi i} F(z) f(z) - \int_{C_2} dz 2\pi i F(z) f(z) - \int_{C_3} dz 2\pi i F(z) f(z).$$

According to Jordan's Lemma the integral over the circle at ∞, C_1, is zero provided that the integrand falls off faster than $1/|z|$.

For example, the occupation of a state, $\mathbf{k}, \sigma$ reads:

$$\langle n_{\mathbf{k}\sigma} \rangle = \frac{1}{\beta} \sum_n \frac{1}{i\omega_n - \epsilon_{\mathbf{k}}} e^{i\omega_n 0^+} = -\int_{C'} \frac{dz}{2\pi i} F(z) f(z) = -\int_{C_2} \frac{dz}{2\pi i} \frac{1}{z - \epsilon_{\mathbf{k}}} e^{z 0^+} f(z) = f(\epsilon_{\mathbf{k}}).$$

Note that the integrand falls off faster than $1/|z|$ and the C_2 contour is shown in Fig. 4.2. Note that C_2 goes in the clockwise direction.

As another important example We can evaluate the Matsubara sum in the bubble diagram:

$$\chi^{(0)}(\mathbf{q}, i\nu_n) = -\frac{2}{\beta V} \sum_{\mathbf{k}, \omega_m} G^{(0)}(\mathbf{k} + \mathbf{q}, i\omega_m + i\nu_n) G^{(0)}(\mathbf{k}, i\omega_m), \tag{4.74}$$

which can be expressed as a contour integral:

$$\chi^{(0)}(\mathbf{q}, i\nu_n) = -\frac{2}{V} \sum_{\mathbf{k}} \int \frac{dz}{2\pi i} F(z) f(z) \tag{4.75}$$

where:

$$F(z) = -\frac{1}{i\nu_n + z - \epsilon_{\mathbf{k}+\mathbf{q}}} \frac{1}{z - \epsilon_{\mathbf{k}}}, \tag{4.76}$$

which has two simple poles at, $z = \epsilon_{\mathbf{k}}$ and $z = \epsilon_{\mathbf{k}+\mathbf{q}} - i\nu_n$. Hence, the deformed C' contour consists now on two circles surrounding these two poles leading to:

$$\chi(\mathbf{q}, i\nu_n) = -\frac{2}{V} \sum_{\mathbf{k}} [G^{(0)}(\mathbf{k}, \epsilon_{\mathbf{k}} - i\nu_n) f(\epsilon_{\mathbf{k}+\mathbf{q}} - i\nu_n) + G^{(0)}(\mathbf{k} + \mathbf{q}, \epsilon_{\mathbf{k}} + i\nu_n) f(\epsilon_{\mathbf{k}})]$$

$$\tag{4.77}$$

This can be expressed as:

$$\chi^{(0)}(\mathbf{q}, i\nu_n) = -\frac{2}{V} \sum_{\mathbf{k}} \left[\frac{f(\epsilon_{\mathbf{k}+\mathbf{q}})}{\epsilon_{\mathbf{k}+\mathbf{q}} - i\nu_n - \epsilon_{\mathbf{k}}} + \frac{f(\epsilon_{\mathbf{k}})}{i\nu_n + \epsilon_{\mathbf{k}} - \epsilon_{\mathbf{k}+\mathbf{q}}} \right] \tag{4.78}$$

where we have used:

$$f(\epsilon_{\mathbf{k+q}} - i\nu_n) = f(\epsilon_{\mathbf{k+q}}). \tag{4.79}$$

Summarizing, the final important result:

$$\chi(\mathbf{q}, i\nu_n) = \frac{2}{V} \sum_{\mathbf{k}} \frac{f(\epsilon_{\mathbf{k+q}}) - f(\epsilon_{\mathbf{k}})}{i\nu_n + \epsilon_{\mathbf{k}} - \epsilon_{\mathbf{k+q}}} \tag{4.80}$$

We will see how this function is relevant in many contexts such as in the response of an electron gas to an external time-dependent potential.

Exercise 1.11 Electron-phonon interaction in solids

Electrons in solids interact with the surrounding lattice through the emission and absorption of "phonons". The interaction between the electrons and phonons is analogous to the electron-photon interaction in QED. The basic hamiltonian describing the interaction of electrons with a vibrating lattice is the Fröhlich hamiltonian:

$$H = \sum_{\mathbf{k},\sigma} \epsilon_{\mathbf{k}} c_{k\sigma}^{\dagger} c_{k\sigma} + \sum_{\mathbf{q},\lambda} \omega_{\mathbf{q}\lambda} \left(b_{\mathbf{q}\lambda}^{\dagger} b_{\mathbf{q}\lambda} + \frac{1}{2}\right)$$

$$+ \sum_{k,\mathbf{q},\lambda} g_{\mathbf{q}\lambda} c_{k+q\sigma}^{\dagger} c_{k\sigma} \left[b_{\mathbf{q}\lambda} + b_{-\mathbf{q}\lambda}^{\dagger}\right]$$

Using finite temperature Feynman diagrams, it is straightforward to obtain the effective interaction between two electrons in a vibrating solid (with no Coulomb interactions between the electrons) at non-zero temperature. Based on the Feynman rules for the electron-phonon interaction graphically determined as:

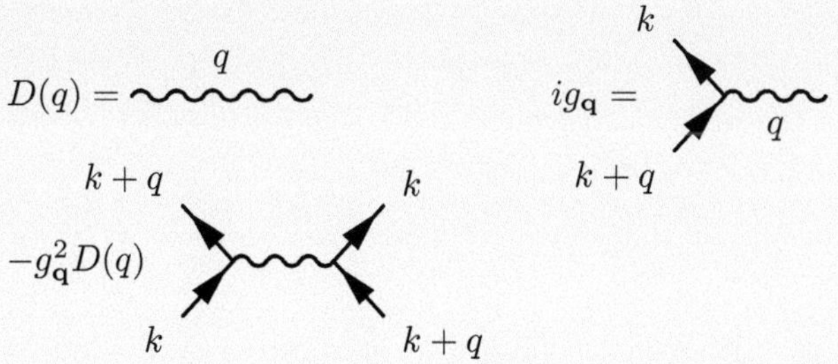

$$D(q) = \qquad\qquad\qquad ig_{\mathbf{q}} = \qquad$$

$$-g_{\mathbf{q}}^2 D(q) \qquad$$

(i) Obtain the expression for the effective interaction, $V_{eff}(\mathbf{q}, \omega)$, between electrons induced by phonon exchange. The effective interaction obtained is retarded since the relaxation time associated with the electron-phonon interaction is related to the characteristic phonon frequency in solids, the Debye frequency, ω_D. Explain what kind of effective interaction, $V_{eff}(\mathbf{q}, \omega)$, arises when the frequency of the effective interaction $\omega < \omega_D$. This low energy component of the interaction is responsible for conventional superconductivity in metals.

(ii) Draw the leading-order Feynman diagram for the self-energy in order to determine the effect of the electron-phonon coupling in electron propagation. Write the expression for the self-energy in terms of Matsubara frequencies.

(iii) Perform the Matsubara sums obtained in (ii) explicitly. Obtain the decay rate, $1/\tau_{\mathbf{k}} = -2 Im \Sigma(\mathbf{k}, \omega + i\eta)$, of an electron due to the electron-phonon interaction from the imaginary part of the self-energy. Give a physical interpretation of the expression obtained.

This exercise is set and solved in [6].

Chapter 5
Linear Response and Collective Modes

Suppose a system described by an interacting hamiltonian, $H = H_0 + V$, perturbed by an external force acting at $t > 0$:

$$\tilde{H} = H - f(t)A = H + \theta(t)V(t), \tag{5.1}$$

where A is any bosonic operator, such as the electron density or spin density. We would like to find the temporal evolution of the average value of A, $A(t)$ in the Heisenberg representation under the effect of the external time-dependent field. Hence, how do we obtain $\langle A_H(t) \rangle$?.

We start by expressing $A_H(t)$ in the interaction representation:

$$A_H(t) = U^\dagger(t, 0)A_I(t)U(t, 0), \tag{5.2}$$

where $U(t, t')$ is the solution to the EOM:

$$i\frac{\partial U(t, t')}{\partial t} = V_I(t)U(t, t') \tag{5.3}$$

which is valid even if the hamiltonian in the Schrödinger picture is explicitly time-dependent. This EOM has the formal solution:

$$U(t, t') = T \exp[-i \int_{t'}^{t} V_I(t_1)dt_1]. \tag{5.4}$$

Expanding to linear order we have:

$$U(t, 0) \approx 1 - i \int_0^t dt_1 V_I(t_1) = 1 + i \int_0^t dt_1 A_I(t_1)f(t_1)$$

$$U^\dagger(t, 0) \approx 1 - i \int_0^t dt_1 A_I(t_1)f(t_1) \tag{5.5}$$

J. Merino and A. L. Yeyati, *Many-Body Techniques in Condensed Matter Physics*, UNITEXT for Physics, https://doi.org/10.1007/978-3-031-55143-7_5

Hence, to linear order the observable, $A_H(t)$:

$$A_H(t) \approx (1 - i \int_0^t dt_1 f(t_1) A_I(t_1)) A_I(t).(1 + i \int_0^t dt_1 A_I(t_1) dt_1)$$

$$= A_I(t) + i \int_0^t dt_1 [A_I(t), A_I(t_1)] f(t_1).$$

By taking thermal averages we find, to linear order:

$$\langle \delta A_H(t) \rangle = \langle A_H(t) - A_I(t) \rangle = i \int_0^{t_1} dt_1 \langle [A_I(t), A_I(t_1)] \rangle f(t_1), \tag{5.6}$$

which is the famous general Kubo formula citetKubo1957. The Kubo formula defines a retarded response function or susceptibility:

$$\langle \delta A_H(t) \rangle = \int_{-\infty}^{\infty} dt_1 \chi^R(t - t_1) f(t_1), \tag{5.7}$$

where:

$$\chi^R(t - t_1) \equiv i \langle [A_I(t), A_I(t_1)] \rangle \theta(t - t_1). \tag{5.8}$$

This is an intrinsic property of the system on which the perturbation is applied.

Note that $A_I(t) = [A_H(t)]_H$ where $[A_H(t)]_H$ refers to the Heisenberg representation of the observable A with respect to the initial fully interacting hamiltonian, H, before plugging the time-dependent perturbation. This follows from:

$$[A_H(t)]_H = e^{iHt} A_S e^{-iHt} = e^{iHt} e^{-iHt} A_I(t) e^{iHt} e^{-iHt} = A_I(t). \tag{5.9}$$

This relation allows to interpret, $\delta A_H(t) = A_H(t) - A_I(t)$, as the modification of the observable, $A(t)$, solely due to the external perturbation acting on H for $t > 0$.

In order to use perturbation theory techniques at finite T of the previous chapter we introduce the imaginary time response function:

$$\chi(\tau - \tau') \equiv \langle T_\tau [A_H(\tau), A_H(\tau')] \rangle, \tag{5.10}$$

Similarly to the relations found between Matsubara and real time Green functions, the frequency dependence of the retarded response function experimentally measured can be obtained by analytical continuation to the real axis:

$$\chi^R(\nu) = \chi(\nu + i\eta). \tag{5.11}$$

Response functions are intimately related to two-particle Greens functions of the system. This allows to obtain information of the correlations from the experimentally measured response of the system as dictated by the fluctuation-dissipation theorem. We will discuss this relation explicitly for the charge response of a system.

5.1 Collective Modes in a Fermi Liquid

The electron gas under Coulomb interaction sustains collective excitations called plasmons. We can explore them by applying an external time-dependent potential which couples to the charge density:

$$\tilde{H} = H - \int f(\mathbf{x}, t) n(\mathbf{x}) d\mathbf{x}, \tag{5.12}$$

where: $f(\mathbf{x}, t) = e^2 \phi_{ext}(\mathbf{x}, t)$ with $\phi_{ext}(\mathbf{x}, t)$ the external scalar potential and $n(\mathbf{x}) = \sum_\sigma \Psi_\sigma^\dagger(\mathbf{x}) \Psi_\sigma(\mathbf{x})$, with $\Psi_\sigma(\mathbf{x})$ the field operator.

Using linear response theory we have seen that the response of the gas to a small applied field is:

$$\langle \delta n(\mathbf{x}, t) \rangle = \int_{-\infty}^{\infty} dt_1 \int d\mathbf{x}_1 \chi^R(\mathbf{x} - \mathbf{x}_1, t - t_1) f(\mathbf{x}_1, t_1), \tag{5.13}$$

where the retarded charge response function or charge susceptibility reads:

$$\chi^R(\mathbf{x} - \mathbf{x}_1, t - t_1) = -i \langle [n(\mathbf{x}, t), n(\mathbf{x}_1, t_1)] \rangle \theta(t - t_1), \tag{5.14}$$

and the associated Matsubara charge response function:

$$\chi(\mathbf{x}, \tau) = \Pi(\mathbf{x}, \tau) = \langle T_\tau[n(\mathbf{x}, \tau) n(\mathbf{0}, 0)] \rangle \equiv \quad \mathrm{x} \bigcirc 0$$

where $x = (\mathbf{x}, \tau)$. We have introduced the particle-hole Green function, $\Pi(\mathbf{x}, \tau)$, discussed previously showing that it is identical to the charge response function $\chi(\mathbf{x}, \tau)$.

Since the finite-T perturbation expansion applies to the particle-hole two particle Green function, $\Pi(\mathbf{x}, \tau)$, we can obtain the terms to any order in the interaction diagrammatically by drawing all topologically inequivalent connected diagrams and evaluating them through the Feynman rules introduced in the previous chapter.

The particle-hole or polarization bubble $\Pi(\mathbf{x}, \tau)$ in momentum space reads:

$$\chi(\mathbf{q}, \tau) = \langle T_\tau[n(\mathbf{q}, \tau) n^\dagger(\mathbf{q}, 0)] \rangle. \tag{5.15}$$

This can be expressed as a Matsubara sum:

$$\Pi(\mathbf{q}, \tau) = \frac{1}{\beta} \sum_n e^{-i\nu_n \tau} \Pi(\mathbf{q}, i\nu_n), \tag{5.16}$$

where the Matsubara components are given by the Fourier transform:

$$\Pi(\mathbf{q}, i\nu_n) = \int_0^\beta d\tau\, e^{i\nu_n \tau} \Pi(\mathbf{q}, \tau). \tag{5.17}$$

The transformation from real to momentum space can be achieved through a Fourier decomposition of the fields:

$$n(\mathbf{x}) = \sum_\sigma \Psi_\sigma^\dagger(\mathbf{x})\Psi_\sigma(\mathbf{x}) = \frac{1}{V}\sum_\sigma \sum_{\mathbf{k},\mathbf{k}'} e^{-i(\mathbf{k}-\mathbf{k}')\cdot\mathbf{x}} c_{\mathbf{k}\sigma}^\dagger c_{\mathbf{k}'\sigma}$$

$$= \frac{1}{V}\sum_{\mathbf{q}} e^{i\mathbf{q}\mathbf{x}} \sum_{\mathbf{k}\sigma} c_{\mathbf{k}\sigma}^\dagger c_{\mathbf{k}+\mathbf{q}\sigma} = \frac{1}{V}\sum_{\mathbf{q}} e^{i\mathbf{q}\cdot\mathbf{x}} n(\mathbf{q}). \tag{5.18}$$

where we have defined: $\mathbf{q} = \mathbf{k}' - \mathbf{k}$. Alternatively at point $\mathbf{x}'$:

$$n(\mathbf{x}') = \frac{1}{V}\sum_{\mathbf{q}'} e^{-i\mathbf{q}'\cdot\mathbf{x}'} \sum_{\mathbf{k}',\sigma} c_{\mathbf{k}'+\mathbf{q}'\sigma}^\dagger c_{\mathbf{k}'\sigma} = \frac{1}{V}\sum_{\mathbf{q}'} e^{-i\mathbf{q}'\cdot\mathbf{x}'} n^\dagger(\mathbf{q}'). \tag{5.19}$$

The charge susceptibility is a two-particle Green function that can be expanded. It has two fermion operators at each vertex instead of just one. $\Pi(\mathbf{q}, \tau)$ is then the sum of all the possible diagrams that connect the two vertices $(n(\mathbf{q}), n^\dagger(\mathbf{q}))$ and involve an arbitrary number of Coulomb repulsion lines. Schematically:

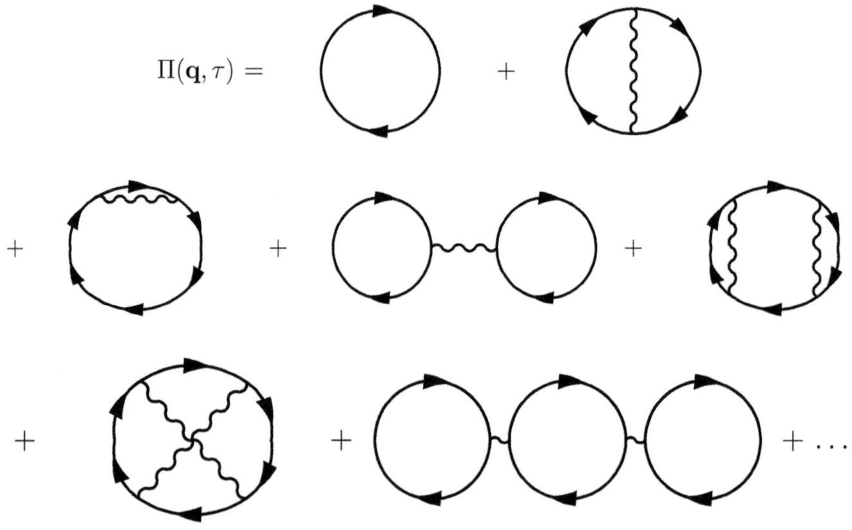

Taking first order diagrams is not a sufficient approximation since Coulomb repulsion V is not small in a metal. The whole series can be expressed in the form of a

Dyson equation in terms of an irreducible $\Pi^{irr}(q)$ which includes the diagrams that cannot be separated in two by cutting a Coulomb line:

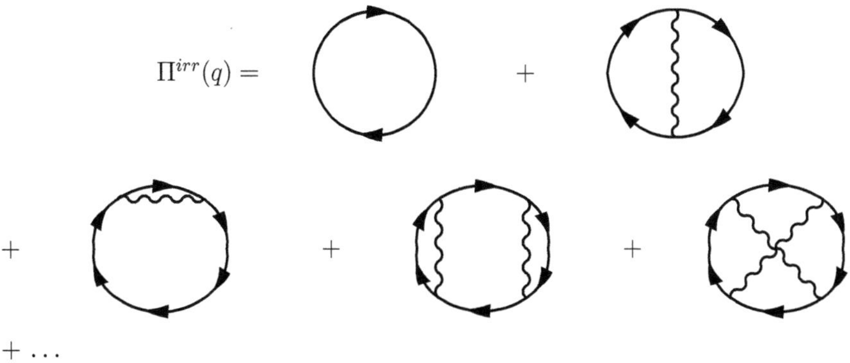

$$\Pi^{irr}(q) =$$

$$+ \quad + \quad +$$

$$+ \, \ldots$$

The full exact polarization $\Pi(q)$ satisfies the Dyson equation:

$$\Pi(q) = \Pi^{irr}(q) - \Pi^{irr}(q)V(q)\Pi(q), \tag{5.20}$$

which can be readily inverted to obtain a formally closed expression for $\Pi(q)$:

$$\Pi(q) = \frac{\Pi^{irr}(q)}{1 + V(q)\Pi^{irr}(\mathbf{q})}, \tag{5.21}$$

We can obtain an approximation describing the electron liquid polarizability or the response to the external field valid at high densities using the RPA to $\Pi(q)$. A diagram of order n can be shown to be proportional to $(k_F^{-2})^n (k_F^{-2})^{2n}$.

5.2 RPA Approximation

It can be shown that the most important contributions to $\Pi(q)$ in the high-density limit ($E_{kin} \gg F_{Coul}$) are given by:

$$\Pi^{irr}(q) \approx \Pi^{(0)}(q) \tag{5.22}$$

where $\Pi^{(0)}(q)$ is the bare particle-hole bubble. The diagrammatic expansion of the particle-hole Green function at the RPA level is:

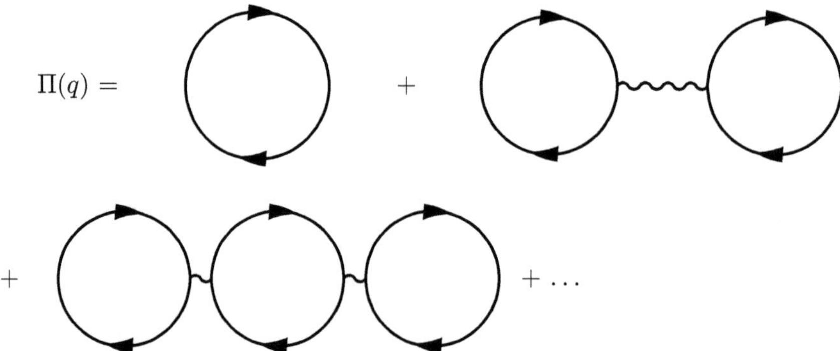

which means summing up all particle-hole bubble contributions to infinite order in the Coulomb interaction.

$$\Pi^{RPA}(q) = \Pi^{(0)}(q) - \Pi^{(0)}(q)V(q)\Pi^{(0)}(q) + \Pi^{(0)}(q)V(q)\Pi^{(0)}(q)V(q)\Pi^{(0)}(q) + \dots \tag{5.23}$$

where the alternating sign comes from the (-1) associated with each Coulomb line as dictated by teh Feynamn rules. Hence, the particle-hole Green function at the RPA level reads:

$$\Pi^{RPA}(q) = \frac{\Pi^{(0)}(q)}{1 + V(q)\Pi^{(0)}(q)}. \tag{5.24}$$

The non-interacting particle-hole density bubble is given by:

$$\Pi^{(0)}(q) = -\frac{2}{\beta}\sum_m \int \frac{d\mathbf{k}}{(2\pi)^3} G^{(0)}(\mathbf{k} + \mathbf{q}, i\omega_m + i\nu_n)G^{(0)}(\mathbf{k}, i\omega_m). \tag{5.25}$$

which was evaluated previously. After performing the Matsubara sum, one finds:

$$\Pi^{(0)}(\mathbf{q}, i\nu_n) = -2\int \frac{d\mathbf{k}}{(2\pi)^3} \frac{f(\epsilon_\mathbf{k}) - f(\epsilon_{\mathbf{k}+\mathbf{q}})}{i\nu_n + \epsilon_\mathbf{k} - \epsilon_{\mathbf{k}+\mathbf{q}}}, \tag{5.26}$$

where $\nu_n = 2\pi n/\beta$. The particle-hole spectral density reads:

$$\frac{1}{\pi}Im\Pi^{(0)R}(\mathbf{q}, \nu) = \frac{1}{\pi}Im\Pi^{(0)}(\mathbf{q}, \nu + i\eta) = \frac{2}{Vol}\sum_\mathbf{k}(f(\epsilon_\mathbf{k}) - f(\epsilon_{\mathbf{k}+\mathbf{q}}))\delta(\nu + \epsilon_\mathbf{k} - \epsilon_{\mathbf{k}+\mathbf{q}}). \tag{5.27}$$

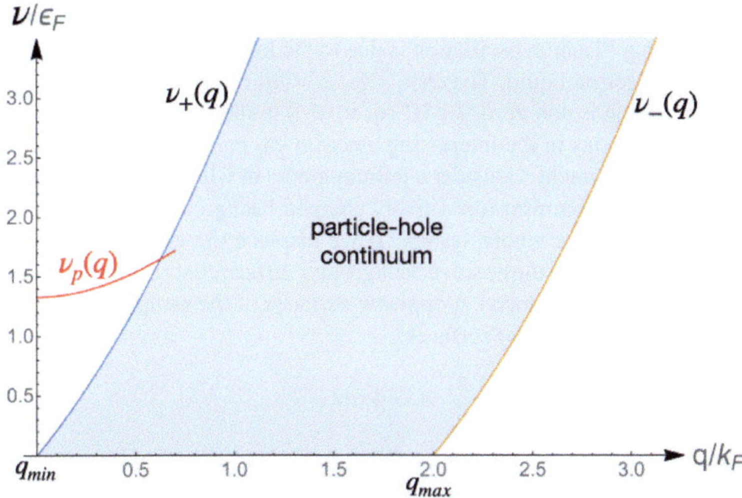

Fig. 5.1 $\omega - q$ region in which the particle-hole spectral density of a 3D interacting Fermi gas is non-zero. The allowed particle-hole continuum of excitations (blue shaded area) and the plasmon dispersion, $\nu_p(q)$ (red solid line) are shown. $\nu_-(q)$ and $\nu_+(q)$ are the minimum and maximum particle-hole excitation energies. The maximum allowed momentum transfer of a particle-hole excitation with $\nu = 0$ is $q_{max} = 2k_F$. Plasmons become unstable on entering the particle-hole continuum since they can decay in particle-hole pairs. In the non-interacting Fermi gas only the particle-hole continuum is allowed since plasmons are a consequence of the long range Coulomb interaction

$Im\,\Pi^{(0)R}(\mathbf{q}, \nu) \neq 0$ responds to the ability of the e^- gas to absorb the external energy and momentum, $(\mathbf{q}, \nu)$ by generating $e - h$ pairs. The $\omega(|\mathbf{q}|)$ vs. $q = |\mathbf{q}|$ plane in which $Im\,\Pi^{(0)R}(\mathbf{q}, \nu) \neq 0$ corresponds to the particle-hole continuum shown in Fig. 5.1.

In contrast, in an interacting electron gas, a new collective excitation arises apart from the particle-hole continuum of excitations. This is the plasmon excitation in which the whole gas oscillates collectively and arises as a pole in the fully dressed charge polarization propagator, $\Pi^R(\mathbf{q}, \nu)$. An analytic expression can be obtained in the $\nu >> v_F|\mathbf{q}|$ limit, where the bare particle-hole bubble is real: $Im\,\Pi^{(0)R}(\mathbf{q}, \nu) = 0$. In this limit:

$$\Pi^{(0)R}(\mathbf{q}, \nu) = -\frac{n}{m}\frac{|\mathbf{q}|^2}{|\nu|^2}. \tag{5.28}$$

Introducing this in the expression for the fully dressed propagator (5.24) one finds:

$$\Pi^R(\mathbf{q}, \nu) = \frac{\Pi^{(0),R}(\mathbf{q}, \nu)}{1 + \frac{4\pi e^2}{|\mathbf{q}|^2}\Pi^{(0),R}(\mathbf{q}, \nu)} = \frac{-\frac{n}{m}\frac{|\mathbf{q}|^2}{\nu^2}}{1 - \frac{4\pi e^2 n}{m\nu^2}}, \tag{5.29}$$

which has a pole when $\nu^2 = \nu_p^2 = \frac{4\pi n e^2}{m}$ associated with the plasmon collective mode with no damping. Plasmon oscillation is due to the long range Coulomb interactions present in the electron liquid. The $\omega(|\mathbf{q}|)$ vs. $q = |\mathbf{q}|$ region where the dressed RPA charge polarization is non-zero, $Im\,\Pi^R(\mathbf{q}, \nu) \neq 0$ is shown in Fig. 5.1.

Plasma oscillations in the interacting electron gas can be understood based on a simple classical argument. Consider a jellium model in which a uniform electron gas is in the presence of a uniform positively charged background (jellium) preserving charge neutrality of the whole system. If we displace the electron gas by a small quantity x with respect to the positive background surface charge densities of opposite sign, $\pm\sigma = \pm nex$, are induced at opposite surfaces of the sample. This leads to an electric field between the two surfaces:

$$\epsilon = 4\pi nex. \tag{5.30}$$

Newton's second law would lead to the equation of motion:

$$m\frac{d^2x}{dt^2} = -e\epsilon = -4\pi ne^2 x \tag{5.31}$$

which is just the equation of an harmonic oscillator:

$$\frac{d^2x}{dt^2} + \frac{4\pi ne^2}{m}x = 0, \tag{5.32}$$

which describes oscillations of the whole electron gas around the equilibrium position with frequency:

$$\omega_p^2 = \frac{4\pi ne^2}{m}. \tag{5.33}$$

5.3 Screening of Coulomb Repulsion

We can also consider the self-energy diagrams at the RPA level:

$$\Sigma_\sigma^{RPA}(q) \quad = \quad$$

where the bare Coulomb interaction between electrons has been renormalized:

Hence, the effective Coulomb interaction itself also satisfies a Dyson equation:

$$V^{eff}(q) = \frac{V(\mathbf{q})}{1 + V(\mathbf{q})\Pi^{(0)}(q)}. \tag{5.34}$$

This is the effective interaction between electrons at the RPA level in actual metals which contains screening effects of the rest of the electrons in the metal. We can gain insight into the structure of the effective Coulomb interaction by analyzing various limits.

In the limit ($\omega << v_F|\mathbf{q}|$) (*i. e.* in the static limit) in which electrons have reached the stationary situation the bare polarization can be approximated as:

$$\Pi^{(0)R}(\mathbf{q}, 0) = -2 \int \frac{d\mathbf{k}}{(2\pi)^3} \frac{\partial f}{\partial \epsilon_{\mathbf{k}}} = 2 \int \frac{d\mathbf{k}}{(2\pi)^3} \frac{\partial f}{\partial \mu} = \frac{\partial n}{\partial \mu}, \tag{5.35}$$

where, $|\mathbf{q}| \to 0$, has been assumed. n is the particle density of the electron gas.

Defining the Thomas-Fermi screening wavelength:

$$\kappa_{TF}^2 = 4\pi e^2 \frac{\partial n}{\partial \mu}, \tag{5.36}$$

we have the effective Coulomb interaction in the static limit:

$$V_{eff}^{RPA}(\mathbf{q}) = \frac{\frac{4\pi e^2}{|\mathbf{q}|^2}}{1 + \frac{4\pi e^2}{|\mathbf{q}|^2} \frac{\partial n_e}{\partial \mu}} = \frac{4\pi e^2}{\kappa_{TF}^2 + |\mathbf{q}|^2}. \tag{5.37}$$

Fourier transforming we find the Yukawa interaction:

$$V_{eff}(\mathbf{r}) = \int \frac{d\mathbf{q}}{(2\pi)^3} e^{i\mathbf{q}\cdot\mathbf{r}} \frac{4\pi e^2}{\kappa_{TF}^2 + |\mathbf{q}|^2} = e^2 \frac{e^{-\kappa_{TF}r}}{r}. \tag{5.38}$$

An electron in a metal carries a spherical screening cloud of extension, λ_{TF} as shown in Fig. 5.2. The effective electrostatic interaction between an electron and the screened electron is precisely given by $V_{eff}(\mathbf{r})$ which at large distances $r > \lambda_{TF}$ is

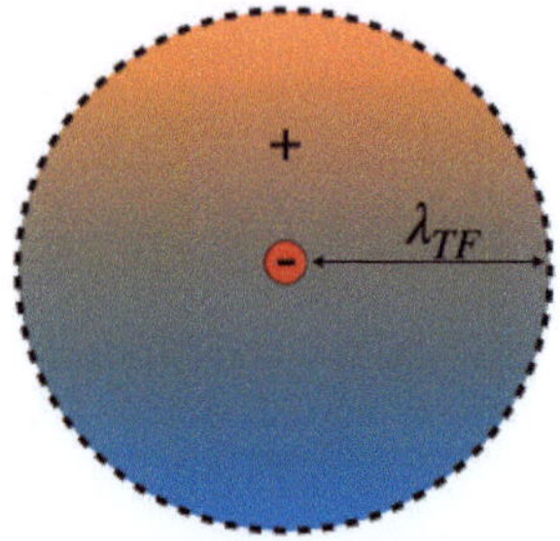

Fig. 5.2 Schematic representation of screening in an interacting 3D Fermi gas. The spatial extension of the positive (hole) cloud around an electron in the gas is λ_{TF} from RPA in the static approximation

screened out. Thus, at these distances (low energies) the system effectively behaves as if it was non-interacting.

An interesting application of the RPA is the exploration of possible charge, magnetic and/or superconducting instabilities in a Fermi liquid citetScalapino1994.

Exercise I.1: Charge order instabilities

Charge ordering phenomena is observed in several strongly correlated materials including cuprate superconductors and layered organic materials (Merino 2006). We consider an extended Hubbard model on the square lattice:

$$H = -t \left(\sum_{\langle ij \rangle} c_{i\sigma}^{\dagger} c_{j\sigma} + c_{j\sigma}^{\dagger} c_{i\sigma} \right) + U \sum_{i} n_{i\uparrow} n_{i\downarrow} + V \sum_{\langle ij \rangle} n_i n_j - \mu \sum_{i\sigma} n_{i\sigma},$$

$$(5.39)$$

where $\langle ij \rangle$ denotes nearest-neighbor sites. U is on-site and V nearest-neighbor Coulomb repulsion.

(i) Write down the expression for the charge response function, $\chi_c(\mathbf{q}, \nu)$, at the RPA level. At half-filling, give the condition on the U and V parameters for a (π, π, π) charge order instability.

(ii) Obtain the analytical expression of the electron self-energy at finite-T given by the Feynman diagram:

$$\Sigma_{\sigma}^{RPA}(q) = $$

in which the double wiggly line is the screened Coulomb interaction treated at the RPA level.

5.4 Screening of an External Potential in the Electron Liquid

The external potential $\phi_{ext}(\mathbf{x}, t)$ generates an induced charge density, $-e\delta n(\mathbf{x}, t)$, in the electron liquid which, in turn, leads to an induced potential:

$$\phi_{ind}(\mathbf{x}, t) = \int d\mathbf{x}` \frac{-e\delta n(\mathbf{x}', t)}{|\mathbf{x} - \mathbf{x}'|} \tag{5.40}$$

which implies:

$$\phi_{ind}(\mathbf{q}, \omega) = \frac{4\pi}{|\mathbf{q}|^2}[-e\delta n(\mathbf{q}, \omega)] \tag{5.41}$$

From the linear response theory introduced previously, the induced electron density reads:

$$-e\delta n(\mathbf{q}, \omega) = -e^2 \chi^R(\mathbf{q}, \omega)\phi_{ext}(\mathbf{q}, \omega), \tag{5.42}$$

which leads to the relation:

$$\phi_{ind}(\mathbf{q}, \omega) = -\frac{4\pi e^2}{|\mathbf{q}|^2} \chi^R(\mathbf{q}, \omega)\phi_{ext}(\mathbf{q}, \omega). \tag{5.43}$$

The total effective potential acting on an electron will then be the sum of the external and induced potentials:

$$\phi_{tot}(\mathbf{q}, \omega) = \phi_{ext}(\mathbf{q}, \omega) + \phi_{ind}(\mathbf{q}, \omega) = (1 - \frac{4\pi e^2}{|\mathbf{q}|^2} \chi^R(\mathbf{q}, \omega))\phi_{ext}(\mathbf{q}, \omega) \tag{5.44}$$

In analogy with electrostatics in dielectric media we have, $\mathbf{D} = \epsilon\mathbf{E}$ where $\mathbf{D}$ (ϕ_{ext}) and $\mathbf{E}$ (ϕ_{tot}), we have the relation:

$$\phi_{tot}(\mathbf{q}, \omega) = \frac{\phi_{ext}(\mathbf{q}, \omega)}{\epsilon(\mathbf{q}, \omega)}, \tag{5.45}$$

with the dielectric constant of the electron liquid:

$$\epsilon(\mathbf{q}, \omega)^{-1} = 1 - \frac{4\pi e^2}{|\mathbf{q}|^2} \chi^R(\mathbf{q}, \omega). \tag{5.46}$$

Introducing the RPA expression for $\chi^R(\mathbf{q}, \omega) = \Pi^R(\mathbf{q}, \omega)$, we have:

$$\epsilon(\mathbf{q}, \omega)^{-1} = 1 - \frac{4\pi e^2}{|\mathbf{q}|^2} \frac{\Pi^{(0)R}(\mathbf{q}, \omega)}{1 + \frac{4\pi e^2}{|\mathbf{q}|^2}\Pi^{(0)R}(\mathbf{q}, \omega)} = \frac{1}{1 + \frac{4\pi e^2}{|\mathbf{q}|^2}\Pi^{(0)R}(\mathbf{q}, \omega)}, \tag{5.47}$$

where $\Pi^{(0)R}(\mathbf{q}, \omega)$ is the bare retarded particle-hole propagator. Hence $\epsilon(\mathbf{q}, \omega)$ at the RPA level reads:

$$\epsilon(\mathbf{q}, i\omega_n) = 1 + \frac{4\pi e^2}{|\mathbf{q}|^2} \Pi^{(0)R}(\mathbf{q}, \omega), \qquad (5.48)$$

so that the total potential inside the electron liquid is:

$$\phi_{tot}(\mathbf{q}, \omega) = \frac{\phi_{ext}(\mathbf{q}, \omega)}{1 + \frac{4\pi e^2}{|\mathbf{q}|^2} \Pi^{(0)R}(\mathbf{q}, \omega)}, \qquad (5.49)$$

so an electron in a Fermi liquid feels the potential screened by the rest of the electrons rather than the bare external potential.

Chapter 6
Spontaneous Symmetry Breaking and Mean-Field Theory

We now consider states of matter which occur, for instance, in phase transitions when temperature is reduced below the critical temperature, $T < T_c$. These states are characterized by having long range order and spontaneous symmetry breaking as in ferromagnets.

As advanced earlier in the course spontaneous symmetry breaking is not described by standard perturbation theory nor Fermi liquid theory. This is because FDPT is based on the adiabatic hypothesis which excludes interacting ground states with different symmetry from the unperturbed non-interacting state around which perturbation theory is applied.

More explicitly spontaneous symmetry breaking means that the perturbation series breaks down since the final ground state is orthogonal to the non-interacting initial state, $\langle \Phi_G | \Psi_G \rangle = 0$. This is because $|\Psi_G\rangle$ and $|\Phi_G\rangle$ have different symmetry. In definite, this means that our initial assumption about the unperturbed non-interacting state being a normal non-interacting Fermi system *i. e.* misses the possibility of an spontaneously broken symmetry state.

However, perturbation theory could be still applied by previously finding possible spontaneously broken symmetry states in the system. Once these states are identified, perturbation theory around such states instead of the homogeneous normal state could be performed. By a straightforward generalization of FDPT, such spontaneous broken symmetry states can also be treated in a convenient way as we will see in this chapter.

6.1 Generalized Green Function Propagator

Consider, for example, a ferromagnetic state occurring below the Curie temperature, $T < T_c$. Such system would display long range order with all the spins aligned in the same direction. Due to the magnetic field generated by the ferromagnetic spins, an electron can emerge with the opposite spin to the one it had when added. Obviously, this spin flip process when adding an electron was not considered in

© The Author(s), under exclusive license to Springer Nature Switzerland AG 2024

J. Merino and A. L. Yeyati, *Many-Body Techniques in Condensed Matter Physics*,

UNITEXT for Physics, https://doi.org/10.1007/978-3-031-55143-7_6

normal metallic (paramagnetic) phases. Hence, propagators must be extended to deal with such possibility.

We introduce anomalous propagators which include spin-flip processes:

$$G_{\sigma\sigma'}(\mathbf{k}, t - t') = -i\langle\Psi_G|T[c_{\mathbf{k}\sigma}(t)c^{\dagger}_{\mathbf{k}'\sigma'}(t')]|\Psi_G\rangle. \tag{6.1}$$

Since $\sigma =\uparrow, \downarrow$, this is described by a 2×2 matrix:

$$G_{\sigma\sigma'}(\mathbf{k}, t - t') = -i\begin{pmatrix} \langle\Psi_G|T[c_{\mathbf{k}\uparrow}c^{\dagger}_{\mathbf{k}'\uparrow}]|\Psi_G\rangle & \langle\Psi_G|T[c_{\mathbf{k}\uparrow}c^{\dagger}_{\mathbf{k}'\downarrow}]|\Psi_G\rangle \\ \langle\Psi_G|T[c_{\mathbf{k}\downarrow}c^{\dagger}_{\mathbf{k}'\uparrow}]|\Psi_G\rangle & \langle\Psi_G|T[c_{\mathbf{k}\downarrow}c^{\dagger}_{\mathbf{k}'\downarrow}]|\Psi_G\rangle \end{pmatrix} \tag{6.2}$$

which can be compactly written in terms of spinors:

$$\gamma_{\mathbf{k}}(t) = \begin{pmatrix} c_{\mathbf{k}\uparrow}(t) \\ c_{\mathbf{k}\downarrow}(t) \end{pmatrix} \tag{6.3}$$

as:

$$G(\mathbf{k}, t - t') = -i\langle\Psi_G|T[\gamma_{\mathbf{k}}(t)\gamma_{\mathbf{k}}(t')]|\Psi_G\rangle. \tag{6.4}$$

The hamiltonian of an interacting electron gas can be expressed in terms of these spinors as:

$$H = \sum_{\mathbf{k}} \epsilon_{\mathbf{k}}\gamma^{\dagger}_{\mathbf{k}}\gamma_{\mathbf{k}} + \frac{1}{2} \sum_{\mathbf{k}_1,\mathbf{k}_2,\mathbf{k}_3,\mathbf{k}_4} V_{\mathbf{k}_1,\mathbf{k}_2,\mathbf{k}_3,\mathbf{k}_4} \gamma^{\dagger}_{\mathbf{k}_2}\gamma_{\mathbf{k}_4}\gamma^{\dagger}_{\mathbf{k}_1}\gamma_{\mathbf{k}_3}. \tag{6.5}$$

This means that both G and H have the same form in terms of $\gamma_{\mathbf{k}}$ than the normal G and H in terms of the $c_{\mathbf{k}}$. Hence, it is plausible that we can use the same perturbation expansion as previously with the directed lines now associated with the new matrix propagators instead of the conventional ones. hence, we can still apply the FDPT on the new anomalous G' which would then satisfy a Dyson equation in matrix form:

$$G = (G_0^{-1} - \Sigma[G])^{-1} \tag{6.6}$$

where now 2×2 matrix inversions should be understood in the equation above. Note that $\Sigma[G]$ and not $\Sigma[G_0]$ should be used above, otherwise we would never allow for broken symmetry solutions. Another way to derive these expressions is to turn on a infinitesimally small source field (which is taken to 0 at the end of the calculation) so that the new unperturbed state $|\Phi_G\rangle$ is no longer orthogonal to the spontaneously broken symmetry ground state of the interacting system, $|\Psi_G\rangle$.

6.2 Application to Metallic Ferromagnetism

Let's consider a Fermi gas under a contact electron-electron interaction:

$$V(\mathbf{x}) = U\delta(\mathbf{x}).$$

$$(6.7)$$

which leads to the hamiltonian:

$$H = \sum_{\mathbf{k}\sigma} \epsilon_{\mathbf{k}} c_{\mathbf{k}\sigma}^{\dagger} c_{\mathbf{k}\sigma} + U \sum_{\mathbf{k},\mathbf{k'},\mathbf{q},\sigma,\sigma'} c_{\mathbf{k}+\mathbf{q}\sigma}^{\dagger} c_{\mathbf{k'}-\mathbf{q}\sigma'}^{\dagger} c_{\mathbf{k'}\sigma'} c_{\mathbf{k}\sigma}.$$

$$(6.8)$$

We perform a self-consistent Hartree-Fock approximation in which we assume that the magnetization (if it is eventually favored) is in the z-direction. Hence, the spin-dependent Hartree-Fock self-energy reads:

$$\Sigma^{HF} = \begin{pmatrix} U(\langle n_\uparrow \rangle + \langle n_\downarrow \rangle) - U\langle n_\uparrow \rangle & 0 \\ 0 & U(\langle n_\uparrow \rangle + \langle n_\downarrow \rangle) - U\langle n_\downarrow \rangle \end{pmatrix}$$

$$(6.9)$$

where the longitudinal magnetization is zero: $\Sigma_{\uparrow\downarrow} = \Sigma_{\downarrow\uparrow}$.

The Dyson equation reads:

$$\begin{pmatrix} G_{\uparrow\uparrow}(k) & 0 \\ 0 & G_{\downarrow\downarrow}(k) \end{pmatrix} = \begin{pmatrix} G_{\uparrow\uparrow}^{(0)}(k) & 0 \\ 0 & G_{\downarrow\downarrow}^{(0)}(k) \end{pmatrix}$$

$$+ \begin{pmatrix} G_{\uparrow\uparrow}^{(0)}(k) & 0 \\ 0 & G_{\downarrow\downarrow}^{(0)}(k) \end{pmatrix} \begin{pmatrix} \Sigma_{\uparrow\uparrow}(k) & 0 \\ 0 & \Sigma_{\downarrow\downarrow}(k) \end{pmatrix} \begin{pmatrix} G_{\uparrow\uparrow}(k) & 0 \\ 0 & G_{\downarrow\downarrow}(k) \end{pmatrix}$$

or equating for G:

$$G = (G^{(0)-1} - \Sigma^{HF})^{-1} = \begin{pmatrix} i\omega_n - \epsilon_{\mathbf{k}} - U(\langle n \rangle - \langle n_\uparrow \rangle) & 0 \\ 0 & i\omega_n - \epsilon_{\mathbf{k}} - U(\langle n \rangle - \langle n_\downarrow \rangle) \end{pmatrix}^{-1}$$

which should be understood as a matrix equation.

Summarizing, the HF spin-dependent Green function reads:

$$G_{\sigma\sigma'}(\mathbf{k}, i\omega_n) = \frac{\delta_{\sigma\sigma'}}{i\omega_n - \epsilon_{\mathbf{k}} - U\langle n_{\bar{\sigma}} \rangle}$$

$$(6.10)$$

which is a self-consistent equation The spin-dependent occupations would read:

$$\langle n_\sigma \rangle = \frac{1}{\beta} \sum_n \int \frac{d\mathbf{k}}{(2\pi)^3} G_{\sigma\sigma}(\mathbf{k}, i\omega_n) e^{i\omega_n 0^+} = \int \frac{d\mathbf{k}}{(2\pi)^3} f(\epsilon_{\mathbf{k}} + U\langle n_{\bar{\sigma}} \rangle),$$

$$(6.11)$$

at any T.

We can now express this SCE in terms of the magnetization and total density:

$$m = \langle n_\uparrow \rangle - \langle n_\downarrow \rangle \tag{6.12}$$

$$n = \langle n_\uparrow \rangle - \langle n_\downarrow \rangle \tag{6.13}$$

and so: $\langle n_\uparrow \rangle = \frac{m+n}{2}$ and $\langle n_\downarrow \rangle = \frac{n-m}{2}$. Thus the SCE expressed for m read:

$$m = \int \frac{d\mathbf{k}}{(2\pi)^3} \left(f\left(\epsilon_\mathbf{k} - \frac{Um}{2}\right) - f\left(\epsilon_\mathbf{k} + \frac{Um}{2}\right) \right) \tag{6.14}$$

$$n = \int \frac{d\mathbf{k}}{(2\pi)^3} \left(f\left(\epsilon_\mathbf{k} - \frac{Um}{2}\right) + f\left(\epsilon_\mathbf{k} + \frac{Um}{2}\right) \right) \tag{6.15}$$

which can be expressed in terms of the non-interacting DOS, $D_\sigma(\epsilon) = D(\epsilon)/2$.

$$m = \frac{1}{2} \int d\epsilon D(\epsilon) \left(f\left(\epsilon - \frac{Um}{2}\right) - f\left(\epsilon_\mathbf{k} + \frac{Um}{2}\right) \right) \tag{6.16}$$

$$n = \frac{1}{2} \int d\epsilon D(\epsilon) \left(f\left(\epsilon - \frac{Um}{2}\right) + f\left(\epsilon + \frac{Um}{2}\right) \right) \tag{6.17}$$

which in the limit, $m \to 0^+$ simplifies to:

$$m \approx \frac{Um}{2} \int d\epsilon D(\epsilon) \left(-\frac{\partial f}{\partial \epsilon} \right). \tag{6.18}$$

Thus we find:

$$1 = U \int d\epsilon D_\sigma(\epsilon) \left(-\frac{\partial f}{\partial \epsilon} \right), \tag{6.19}$$

which is known as the Stoner criterion for FM in metals. In the $T \to 0$ limit:

$$1 = U D_\sigma(\epsilon_F), \tag{6.20}$$

where we have used that for $T \to 0$

$$\left(-\frac{\partial f}{\partial \epsilon} \right) = \delta(\epsilon - \epsilon_F). \tag{6.21}$$

6.3 Elementary Excitations in a Broken Symmetry Phase

We may now ask ourselves how do excitations in a broken symmetry phase such as
the FM phase above look like? Apart from the expected particle-hole excitations in
the FM state we should have collective modes: spin waves. These are the Goldstone

modes associated with breaking of a continuous rotation symmetry of the spins (since a particular direction is selected by all the spins of the lattice although any spin direction leads to the same energy).As such spin waves in ferromagnets are gapless, since a $\mathbf{q} \to 0$ uniform rotation of all spins would cost no energy.

Neutron scattering experiments can probe magnetic excitations since neutron spins couple with the electron spin. Hence, instead of measuring charge correlations, neutrons measure spin correlations, $S(\mathbf{q}, \omega) = \int e^{i(\mathbf{qx}-\omega t)} \langle S(\mathbf{x}, t) S(\mathbf{0}, 0) \rangle$. As we have seen these correlations can be related to its corresponding response function through the fluctuation-dissipation theorem:

$$S(\mathbf{q}, \omega) \propto -Im\chi^{R+-}(\mathbf{q}, \omega).$$

where:

$$\chi^{R+-}(\mathbf{q}, t) = i \langle [S^+(\mathbf{q}, t) S^-(\mathbf{q}, t)] \rangle \theta(t). \tag{6.22}$$

The spin operators can be expressed in terms of fermion operators are:

$$S_i^\alpha = \frac{1}{2} \sum_{\sigma\sigma'} c_{i\sigma}^\dagger \sigma_{\sigma\sigma'}^\alpha c_{i\sigma'}, \tag{6.23}$$

where σ^α are the Pauli matrices.

We want to explore spin excitations around the FM broken symmetry state in the interacting Fermi liquid from the evaluation of the two-particle spin response function, $\chi^{+-}(\mathbf{q}, i\omega_n)$. The starting point is to introduce the bare two-particle spin propagators in terms of the generalized HF Green functions obtained previously:

$$\chi^{(0)+-}(q) = -\frac{1}{\beta} \sum_m \int \frac{d\mathbf{k}}{(2\pi)^3} Tr[\sigma^+ G^{HF}(\mathbf{k}+\mathbf{q}, i\omega_m + i\nu_n)\sigma^- G^{HF}(\mathbf{k}, i\omega_m)], \tag{6.24}$$

and:

$$G^{HF}(\mathbf{k}, i\omega_n) = \begin{pmatrix} \frac{1}{i\omega_n + \tilde{\epsilon}_\mathbf{k} - \frac{mU}{2}} & 0 \\ 0 & \frac{1}{i\omega_n + \tilde{\epsilon}_\mathbf{k} + \frac{mU}{2}} \end{pmatrix} \tag{6.25}$$

The Coulomb interaction is treated at the RPA level:

$$\chi^{+-}(q) = \frac{\chi^{(0)+-}(q)}{1 - U\chi^{(0)+-}(q)}, \tag{6.26}$$

with $q = (\mathbf{q}, \nu_n)$, where:

$$\chi^{(0)+-}(q) = \int \frac{d\mathbf{k}}{(2\pi)^3} \frac{f(\tilde{\epsilon}_{\mathbf{k}+\mathbf{q},\downarrow}) - f(\tilde{\epsilon}_{\mathbf{k},\uparrow})}{i\nu_n + \tilde{\epsilon}_{\mathbf{k},\uparrow} - \tilde{\epsilon}_{\mathbf{k}+\mathbf{q},\downarrow}} \tag{6.27}$$

with the spin-dependent energies:

$$\tilde{\epsilon}_{k\sigma} = \epsilon_k + \bar{\sigma} U m/2.$$
$$\tilde{\epsilon}_{k,\uparrow} - \tilde{\epsilon}_{k+q,\downarrow} = \Delta \tag{6.28}$$

where $\Delta = Um$ is the gap between spin-up and spin-down Fermi surfaces.

On the other hand, the bare longitudinal response function reads:

$$\chi^{(0)zz}(q) = \int \frac{d\mathbf{k}}{(2\pi)^3} \frac{f(\epsilon_{k+q}) - f(\epsilon_k)}{i\nu_n + \epsilon_k - \epsilon_{k+q}} \tag{6.29}$$

which coincides with the charge correlation function.

We now analyze the magnetic correlations of a metallic ferromagnet measured in neutron scattering. Based on the fluctuation-dissipation theorem the response function determining the magnegtic escitations reads:

$$-\frac{1}{\pi} Im\chi^{R+-}(\mathbf{q}, \nu) = -\frac{1}{\pi} Im\chi^{+-}(\mathbf{q}, \nu + i\eta), \tag{6.30}$$

which describes the particle-hole excitations involving a spin-flip process.

For $\mathbf{q} = 0$, only a particle-hole pair with energy: $\nu = \Delta$ can be generated since $\tilde{\epsilon}_{k,\uparrow} - \tilde{\epsilon}_{k,\downarrow} = \Delta$, and corresponds to a transition from the small spin down Fermi surface to the large spin-up Fermi surface.

For $\omega = 0$, a minimum (maximum) momentum of $q_{min} = k_{F\uparrow} - k_{F\downarrow}(q_{max} = k_{F\uparrow} + k_{F\downarrow})$ is needed to generate particle-hole pairs from the small spin down Fermi surface up to the large spin-up Fermi surface.

The full region of particle-hole excitations is shown in Fig, 6.1.

Similarly to plasmon modes encountered in the charge response functions, magnetic collective mode, s i.e. spin waves, emerge in the magnetic response function when:

$$1 - U\chi^{(0)+-}(\mathbf{q}, \nu) = 0. \tag{6.31}$$

which leads to the dispersion:

$$\nu(\mathbf{q}) \propto |\mathbf{q}|^2. \tag{6.32}$$

This is a consequence of Goldstone theorem by which a collective mode with a dispersion, $\nu_G(q)$, that goes to zero as $|\mathbf{q}| \to 0$ emerges within the broken symmetry ground state. The spin wave is well defined up to a critical $\mathbf{q}_c$ at which it merges with the continuum of Stoner excitations. For $q > q_c$ the spin waves can decay through the formation of particle-hole Stoner excitations.

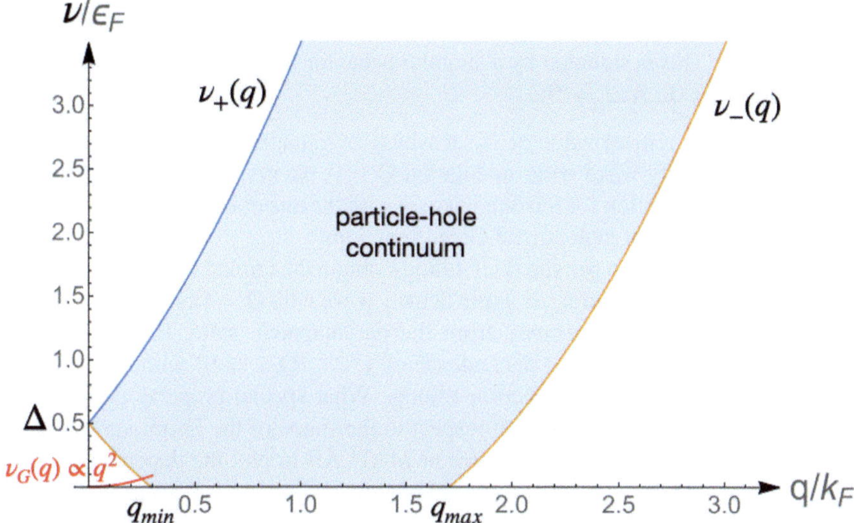

Fig. 6.1 Region of $\omega - q$ in which the particle-hole spectral density in a 3D ferromagnetic Fermi liquid is non-zero. The allowed particle-hole continuum of excitations (blue shaded area) and the Goldstone mode, $\nu_G(q) \propto q^2$ (red solid line) are shown. Here, $q_{min} = k_{F\uparrow} - k_{F\downarrow}$ while $q_{max} = k_{F\uparrow} + k_{F\downarrow}$. We took $\Delta = 0.5\epsilon_F$ and $r_s = 2a_0$

Exercise 1.3: Magnetic instabilities of the Hubbard model

A material can be described by a Hubbard model on a simple cubic lattice. The Hubbard model in momentum space reads:

$$H = \sum_{\mathbf{k}\sigma} \epsilon_{\mathbf{k}} c^{\dagger}_{\mathbf{k}\sigma} c_{\mathbf{k}\sigma} + \frac{U}{N} \sum_{\mathbf{k},\mathbf{k}'\mathbf{q}} c^{\dagger}_{\mathbf{k}+\mathbf{q}\uparrow} c^{\dagger}_{\mathbf{k}'-\mathbf{q}\downarrow} c_{\mathbf{k}'\downarrow} c_{\mathbf{k}\uparrow}$$

where: $\epsilon_{\mathbf{k}} = -2t(\cos(k_x a) + \cos(k_y a) + \cos(k_z a))$, with a the lattice parameter. We are interested in neutron scattering experiments which probe the spin propagator:

$$\chi^{R+-}(\mathbf{q}, t) = i \langle [S^+(\mathbf{q}, t), S^-(\mathbf{q}, 0)] \rangle \theta(t),$$

where $S^+(\mathbf{q}) = \sum_{\mathbf{k}} c^{\dagger}_{\mathbf{k}+\mathbf{q}\uparrow} c_{\mathbf{k}\downarrow}$ and $S^-(\mathbf{q}) = \sum_{\mathbf{k}} c^{\dagger}_{\mathbf{k}+\mathbf{q}\downarrow} c_{\mathbf{k}\uparrow}$ are spin flip operators. At the RPA level the transverse susceptibility in $q = (\mathbf{q}, i\nu_n)$ is given by:

$$\chi^{+-}(q) = \frac{\chi^{(0)+-}(q)}{1 - U\chi^{(0)+-}(q)}.$$

The paramagnetic system can order magnetically spontaneously as U is increased. This is signaled by a singular behavior of the static susceptibility $\chi^{+-}(\mathbf{q}, \nu = 0)$. At $T = 0$:

(i) Find the critical value of U_c at which ferromagnetism, *i. e.* a uniform spin density wave with modulation $\mathbf{Q} = 0$, occurs. Compare the result for the condition for ferromagnetism with the one obtained in class using self-consistent generalized Green functions.

(ii) For one electron per site (half-filling), obtain the critical U at which Néel antiferromagnetism, *i. e.* a spin density wave with $\mathbf{Q} = (\pi/a, \pi/a, \pi/a)$, can occur spontaneously from the paramagnetic state. Explain your results in terms of the dependence of $\chi^{(0)+-}(\mathbf{Q}, \nu = 0)$ with the Fermi energy for different electron fillings. What special property does this spin modulation have with respect to the shape of the Fermi surface at half-filling? (Use Mathematica or MATLAB to plot the dependence of $\chi^{(0)+-}(\mathbf{Q}, \nu = 0)$ with the Fermi energy to explain your results.)

Part II
Non-equilibrium Many-Body Techniques

Chapter 7
Introduction to Non-equilibrium: The Keldysh Contour

In the first part of this course we have addressed many-body techniques for equilibrium systems, either at zero or finite temperature. However, most situations of actual experimental interest correspond to non-equilibrium situations. This is due to the fact that most ways to test the properties of condensed matter systems require interaction with external fields, like in photo-emission or transport experiments, which leads to non-equilibrium quasiparticle distributions. In addition, there is at present great interest in driven systems where non-equilibrium can induce new type of phases, including superconducting or topological phases (Cavalleri 2018; Rudner and Lindner 2020).

Such problems can be generically described by a time-dependent many-body Hamiltonian of the form $H(t) = H_0 + V(t)$ where, in contrast to the assumptions made in the first part of the course, the interaction term is not adiabatically switched off at long times. Let us recall that such assumption was the basis of the "adiabatic hypothesis" connecting the system ground state in the remote past $|\Phi(t \to -\infty)\rangle$ with the state in the remote future $|\Phi(t \to +\infty)\rangle$ by a simple phase factor, i.e. $|\Phi(t \to +\infty)\rangle = e^{i\alpha}|\Phi(t \to +\infty)\rangle$ (Gell-Mann-Low theorem).

In order to address a general non-equilibrium situation it is necessary to "sacrifice" such adiabatic hypothesis and allow a general time-dependence at long times. The adiabatic switch on in the remote past can, in contrast, be kept. A key quantity to be determined is the system density matrix $\rho(t)$ which satisfy the von-Newmann equation,

$$\partial_t \rho(t) = -i\left[H(t), \rho(t)\right]. \tag{7.1}$$

The knowledge of $\rho(t)$ then allows us to obtain the mean value of any observable $\mathcal{O}$ as $\langle\mathcal{O}\rangle(t) = \mathrm{Tr}\left[\mathcal{O}\rho(t)\right]/\mathrm{Tr}\left[\rho\right]$. Formally, Eq. (7.1) can be solved in terms on the evolution operator $\mathcal{U}(t, t')$ in the Schrödinger representation, by means of which we can write

$$\rho(t) = \mathcal{U}(t, -\infty)\rho(-\infty)\mathcal{U}(-\infty, t), \tag{7.2}$$

where $\rho(-\infty)$ is an initial condition which can be taken to correspond to an equilibrium situation for H_0, i.e. $\rho(-\infty) \equiv \exp\left[-\beta(H_0 - \mu N)\right]$. This last assumption,

© The Author(s), under exclusive license to Springer Nature Switzerland AG 2024   97
J. Merino and A. L. Yeyati, *Many-Body Techniques in Condensed Matter Physics*,
UNITEXT for Physics, https://doi.org/10.1007/978-3-031-55143-7_7

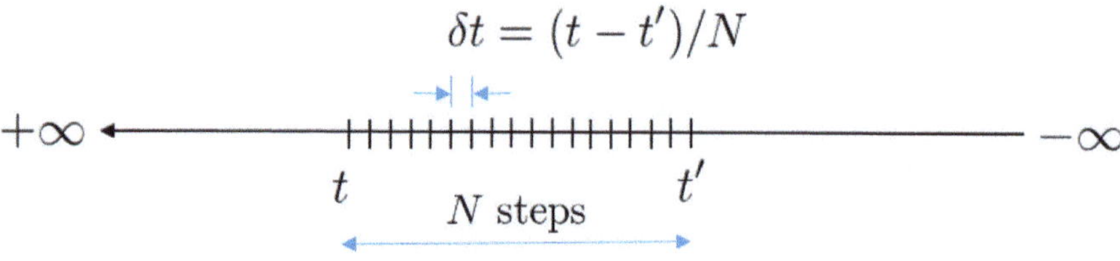

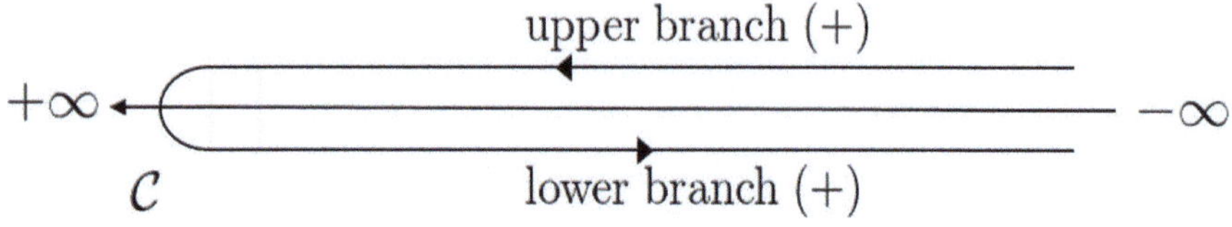

Fig. 7.1 Upper panel: discretization of the real time axis. Lower panel: The Keldysh contour $\mathcal{C}$

although simple, is not completely necessary as in most cases of interest the system steady state can be proven to be independent of the initial condition for ρ (Stefanucci and van Leeuwen 2013).

Let us comment a few subtleties that appear when a perturbative expansion of $\mathcal{U}(t, t')$ is attempted. First notice that $\mathcal{U}(t, t')$ satisfies

$$\partial_t \mathcal{U}(t, t') = -i H(t) \mathcal{U}(t, t')$$
$$\partial'_t \mathcal{U}(t, t') = i \mathcal{U}(t, t') H(t'). \tag{7.3}$$

Since, in general, $\left[H(t), H(t')\right] \neq 0$, $\mathcal{U}(t, t') \neq \exp\left[-i \int_{t'}^{t} H(t_1) dt_1\right]$. In contrast, for an infinitesimal time evolution

$$\mathcal{U}(t' + \delta t, t') \simeq e^{-i \delta t H(t')}.$$

We can thus discretize the time axis as shown in Fig. 7.1 and express $\mathcal{U}(t, t')$ as

$$\mathcal{U}(t, t') = \lim_{N \to \infty} e^{-i \delta t H(t - \delta t)} \ldots e^{-i \delta t H(t')}, \tag{7.4}$$

where $\delta t = (t - t')/N$, which can be formally expressed as

$$\mathcal{U}(t, t') = \mathbf{T} \exp\left[-i \int_{t'}^{t} H(t_1) dt_1\right]. \tag{7.5}$$

In this expression $\mathbf{T}$ is the usual chronological operator which indicates time ordering in the causal way (already discussed in the first part of the course).

Coming back to the evaluation of mean values, we would thus have

$$\langle \mathcal{O} \rangle(t) = \frac{\text{Tr} \left[\overbrace{\mathcal{U}(t, -\infty)}^{\text{forward}} \rho(-\infty) \overbrace{\mathcal{U}(-\infty, t)}^{\text{backward}} \mathcal{O} \right]}{\text{Tr}\,[\rho]} ; \qquad (7.6)$$

which clearly involves a forward ($\mathcal{U}(t, -\infty)$) and a backward ($\mathcal{U}(-\infty, t)$) evolution. Notice that for the backward case with $t > t'$

$$\mathcal{U}(t', t) \neq \mathbf{T} \exp\left[-i \int_t^{t'} H(t_1) dt_1 \right], \qquad (7.7)$$

but rather

$$\mathcal{U}(t', t) = \mathcal{U}(t, t')^\dagger = \lim_{N \to \infty} e^{i\delta t H(t')} \ldots e^{i\delta t H(t - \delta t)}$$

$$\equiv \tilde{\mathbf{T}} \exp\left[-i \int_t^{t'} H(t_1) dt_1 \right], \qquad (7.8)$$

where $\tilde{\mathbf{T}}$ denotes *anticausal* chronological order.

In the equilibrium case one can get rid of the backward part by invoking the adiabatic hypothesis. Let us remind this procedure for the zero temperature case. Let $|0\rangle$ correspond to the ground state of H_0. Then the adiabatic hypothesis means $\mathcal{U}(\infty, -\infty)|0\rangle = e^{i\alpha}|0\rangle$. Therefore,

$$\begin{aligned}
\langle GS | \mathcal{O} | GS \rangle(t) &= \langle 0 | \mathcal{U}(-\infty, t) \mathcal{O} \mathcal{U}(t, -\infty) | 0 \rangle \\
&= \langle 0 | \mathcal{U}(-\infty, \infty) \mathcal{U}(\infty, t) \mathcal{O} \mathcal{U}(t, -\infty) | 0 \rangle \\
&= e^{-i\alpha} \langle 0 | \mathcal{U}(\infty, t) \mathcal{O} \mathcal{U}(t, -\infty) | 0 \rangle \\
&= \frac{\langle 0 | \mathcal{U}(\infty, t) \mathcal{O} \mathcal{U}(t, -\infty) | 0 \rangle}{\langle 0 | \mathcal{U}(\infty, -\infty) | 0 \rangle} ,
\end{aligned} \qquad (7.9)$$

where $|GS\rangle$ denotes the full system ground state. Notice that in this expression only forward evolution appears in the numerator but at the expense of the appearance of a denominator related to the phase factor in the adiabatic hypothesis. As seen in the first part of the course, this denominator leads to the cancellation of disconnected diagrams in the perturbative expansion.

Coming back to the general non-equilibrium situation, it is possible to write the mean values as

$$\begin{aligned}
\langle \mathcal{O} \rangle(t) &= \frac{\text{Tr}\,[\mathcal{U}(-\infty, t) \mathcal{O} \mathcal{U}(t, -\infty) \rho(-\infty)]}{\text{Tr}\,[\rho]} \\
&= \frac{\text{Tr}\,[\mathcal{U}(-\infty, \infty) \mathcal{U}(\infty, t) \mathcal{O} \mathcal{U}(t, -\infty) \rho(-\infty)]}{\text{Tr}\,[\rho]} ,
\end{aligned} \qquad (7.10)$$

which suggests ordering on the close time contour in Fig. 7.1, that we call $\mathcal{C}$ or Keldysh contour. The meaning of this ordering is that, when expanding the evolution operator, one should keep track of the *branch* where each time argument lies: those coming from the forward evolution should be associated with the $+$ branch and those coming from the backward with the $-$ branch. Ordering on $\mathcal{C}$ for $t > t'$ means causal ordering when both belong to the $\mathcal{C}_+$ branch, anticausal if both belong to the $\mathcal{C}_-$ branch and t "previous" to t' if $t \in \mathcal{C}_+$ and $t' \in \mathcal{C}_-$.

One can define $\mathcal{U}_\mathcal{C} = \mathcal{U}_-(-\infty, \infty)\mathcal{U}_+(\infty, -\infty)$ as the evolution operator along the whole $\mathcal{C}$ contour. This seems apparently useless as in all physical conditions $\mathcal{U}_\mathcal{C} = 1$. However, one can artificially break the symmetry between $\mathcal{C}_+$ and $\mathcal{C}_-$ by adding a *source term* to the system Hamiltonian, i.e. defining $H_\chi^\pm(t) = H(t) \pm \chi(t)\mathcal{O}$ where $\pm$ correspond to each branch in $\mathcal{C}$. Then we would have

$$\mathcal{U}_\mathcal{C}[\chi] = \mathbf{T}_\mathcal{C} \exp\left[-i \int_\mathcal{C} H_\chi(t)dt\right] \neq 1 , \tag{7.11}$$

where $\mathbf{T}_\mathcal{C}$ denotes time ordering over $\mathcal{C}$. It is then possible to define a non-equilibrium *partition function* as $Z[\chi] = \mathrm{Tr}\left[\mathcal{U}_\mathcal{C}[\chi]\rho(-\infty)\right]/\mathrm{Tr}[\rho]$, such that

$$\langle\mathcal{O}\rangle(t) = \frac{i}{2}\frac{\delta Z[\chi]}{\delta\chi(t)}\bigg|_{\chi=0} . \tag{7.12}$$

These ideas will be further exploited when discussing the use of the Keldysh formalism to analyze current fluctuations in nanoscale conductors (i.e. the full counting statistics analysis) and in Sect. 10.2 in connection to the path integral formulation of the theory.

Chapter 8
Perturbative Expansion in the Non-equilibrium Formalism

In Eq. (7.10) all operators are in the Schrödinger representation. However, for analyzing the expansion in terms of a generic perturbation $V(t)$ added to the unperturbed Hamiltonian H_0 it is convenient to switch into the *interaction* representation where

$$\mathcal{O}_I(t) = e^{iH_0 t}\mathcal{O}e^{-iH_0 t}$$

$$\mathcal{U}_\pm^I(t, t') \equiv \mathcal{S}_\pm(t, t') = \mathbf{T}_\pm \exp\left[-i\int_{t'}^t V_I(t_1)dt_1\right], \tag{8.1}$$

where we have introduced the most common notation of $\mathcal{S}(t, t')$ for the evolution operator in the interaction representation, now extended to the $\mathcal{C}$ contour. Then, substituting back in (7.10)

$$\langle\mathcal{O}\rangle(t) = \frac{\mathrm{Tr}\left[\mathcal{S}_-(-\infty, \infty)\mathcal{S}_+(\infty, t)\mathcal{O}_I(t)\mathcal{S}_+(t, -\infty)\rho(-\infty)\right]}{\mathrm{Tr}\left[\rho\right]}$$

$$= \frac{\mathrm{Tr}\left[\mathbf{T}_\mathcal{C}\mathcal{S}_\mathcal{C}(\infty, -\infty)\mathcal{O}_I(t)\rho(-\infty)\right]}{\mathrm{Tr}\left[\rho\right]}. \tag{8.2}$$

The next step is to expand $\mathcal{S}_\mathcal{C}$, which can be formally written as

$$\mathcal{S}_\mathcal{C}(\infty, -\infty) = \sum_n \frac{(-i)^n}{n!}\int_\mathcal{C} dt_1 \ldots \int_\mathcal{C} dt_n \mathbf{T}_\mathcal{C}\left[V_I(t_1)\ldots V_I(t_n)\right]. \tag{8.3}$$

For simplicity let us first consider the zero temperature case where $\rho(-\infty) \equiv |0\rangle\langle 0|$. Then

$$\langle\mathcal{O}\rangle(t) = \langle 0|\mathbf{T}_\mathcal{C}\mathcal{O}_I(t)\mathcal{S}_\mathcal{C}(\infty, -\infty)|0\rangle$$

$$= \sum_n \frac{(-i)^n}{n!}\int_\mathcal{C} dt_1 \ldots \int_\mathcal{C} dt_n \langle 0|\mathbf{T}_\mathcal{C}V_I(t_1)\ldots V_I(t_n)\mathcal{O}_I(t)|0\rangle. \tag{8.4}$$

© The Author(s), under exclusive license to Springer Nature Switzerland AG 2024
J. Merino and A. L. Yeyati, *Many-Body Techniques in Condensed Matter Physics*,
UNITEXT for Physics, https://doi.org/10.1007/978-3-031-55143-7_8

Independently of the model considered, both $V(t)$ and $\mathcal{O}$ could be written in terms of field operators $\Psi(x)$, $\Psi^\dagger(x)$ where x denote all possible degrees of freedom, i.e., $x \equiv (r, t, \sigma)$. Additionally, in a canonical description where total number of particles is conserved, the number of creation and annihilation operators should be balanced at each order in the perturbative series. Therefore, a generic term would adopt the form

$$A_N \equiv \langle 0|\mathbf{T}_{\mathcal{C}}\left(\Psi(x_1)\ldots \Psi(x_N)\Psi^\dagger(x_1')\ldots \Psi^\dagger(x_N')\right)|0\rangle ,$$

where all Ψ's and $\Psi^\dagger$'s are in the interaction representation. Such terms could be reduced to a set of single-particle contractions by virtue of Wick's theorem, whose applicability is warranted as H_0 corresponds to a non-interacting problem. Then

$$A_N = \sum_P (\pm 1)^P \langle 0|\mathbf{T}_{\mathcal{C}}\Psi(x_{1P})\Psi^\dagger(x_{1P}')|0\rangle \ldots \langle 0|\mathbf{T}_{\mathcal{C}}\Psi(x_{NP})\Psi^\dagger(x_{NP}')|0\rangle ,$$

where P indicates a given permutation of the particles coordinates and $\pm$ correspond to bosons and fermions respectively. We thus conclude that a basic contraction in the theory is provided by the quantity

$$G^{(0)}(x, x') = -i\langle 0|\mathbf{T}_{\mathcal{C}}\Psi(x)\Psi^\dagger(x')|0\rangle , \tag{8.5}$$

which defines the unperturbed single particle Green function within the non-equilibrium formalism. More generally, for a generic initial distribution we would have

$$G^{(0)}(x, x') = -i\frac{\text{Tr}\left[\mathbf{T}_{\mathcal{C}}\Psi(x)\Psi^\dagger(x')\rho(-\infty)\right]}{\text{Tr}\left[\rho\right]} \equiv -i\langle \mathbf{T}_{\mathcal{C}}\Psi(x)\Psi^\dagger(x')\rangle . \tag{8.6}$$

As can be observed, this definition very much resembles the definition of the causal Green function in the equilibrium case. However, it should be noted that depending on the location of the time arguments t and t' over $\mathcal{C}$ Eq. (8.6) actually corresponds to four possibilities, which we denote by $G^{(0)\alpha\beta}(x, x')$ with α, β indicating the two branches on $\mathcal{C}$. Explicitly we have

$$G^{(0)++}(x, x') = -i\theta(t - t')\langle\Psi(x)\Psi^\dagger(x')\rangle \mp i\theta(t' - t)\langle\Psi^\dagger(x')\Psi(x)\rangle$$
$$G^{(0)--}(x, x') = \pm i\theta(t - t')\langle\Psi^\dagger(x')\Psi(x)\rangle - i\theta(t' - t)\langle\Psi(x)\Psi^\dagger(x')\rangle$$
$$G^{(0)+-}(x, x') = \mp i\langle\Psi^\dagger(x')\Psi(x)\rangle$$
$$G^{(0)-+}(x, x') = -i\langle\Psi(x)\Psi^\dagger(x')\rangle , \tag{8.7}$$

where $\mp$ corresponds to the Bose/Fermi case. Clearly $G^{(0)++}$ corresponds to the causal unperturbed GF as introduced in the first part of the course, while $G^{(0)--}$ corresponds to its *anticausal* counterpart. On the other hand $G^{(0)+-}$ and $G^{(0)-+}$ are new GFs which are needed in the non-equilibrium case. Let us comment that we are following here the notation in the original work by Keldysh (1964), but translation to other notations in the literature, like in Kamenev's book, is straightforward by noting $G^{++} \equiv G^{\mathbf{T}}$, $G^{--} \equiv G^{\tilde{\mathbf{T}}}$, $G^{+-} \equiv G^{<}$ and $G^{-+} \equiv G^{>}$. These last two ones are customarily referred to as *lesser* and *greater* GFs respectively.

8.1   Basic Properties of Non-equilibrium GFs

As we show below, all single particle observables of interest can be expressed in terms of the *complete* or fully dressed single particle GFs given by

$$G(x, x') = -i \langle \mathbf{T}_{\mathcal{C}} S_{\mathcal{C}}(\infty, -\infty) \Psi(x) \Psi^{\dagger}(x') \rangle \tag{8.8}$$

or, in a more compact way $G(x, x') = -i \langle \mathbf{T}_{\mathcal{C}} \Psi_H(x) \Psi_H^{\dagger}(x') \rangle$, where subindex H denotes Heisenberg representation. This quantity can be decomposed in the same way as $G^{(0)}(x, x')$ into its $G^{\alpha\beta}(x, x')$ components, which satisfy Eq. (8.7) but with field operators in the H representation.

It is straightforward to notice that not all $G^{\alpha\beta}$ are independent but are rather linked by the relation

$$G^{++} + G^{--} = G^{+-} + G^{-+}, \tag{8.9}$$

which is satisfied both for the Bose or Fermi case and both for the unperturbed or the fully dressed GFs.[1] Furthermore, the Keldysh GFs can be related to the retarded, advanced GFs defined as

$$\begin{aligned}
G^R(x, x') &= -i\theta(t - t') \langle \left[\Psi(x), \Psi^{\dagger}(x') \right]_{\mp} \rangle \\
G^A(x, x') &= i\theta(t' - t) \langle \left[\Psi(x), \Psi^{\dagger}(x') \right]_{\mp} \rangle,
\end{aligned} \tag{8.10}$$

so that

$$\begin{aligned}
G^R(x, x') &= \theta(t - t') \left(G^{-+}(x, x') - G^{+-}(x, x') \right) \\
G^A(x, x') &= \theta(t' - t) \left(G^{+-}(x, x') - G^{-+}(x, x') \right)
\end{aligned} \tag{8.11}$$

---

[1] Notice, however, that such relation is not satisfied in the presence of a source term breaking the branch symmetry along $\mathcal{C}$.

Exercise II.1: Single level GFs

Consider a simple single quantum level model $H = \epsilon a^\dagger a$; where $a^\dagger, a$ are creation annihilation operators, which can be either fermionic or bosonic. Assume that the initial state corresponds to a thermal equilibrium state with a density matrix $\rho_{eq} = e^{-\beta(H-\mu N)}$.

For the fermionic case:

(i) Obtain the Keldysh single particle Green functions

$$G^{\alpha,\beta}(t,t') = -i\langle \mathbf{T}_{\mathcal{C}} a(t_\alpha) a^\dagger(t'_\beta)\rangle$$

where $\alpha, \beta \equiv +, -$ and average is taken with respect to the initial density matrix.

(ii) For the same model obtain the retarded and advanced GFs

$$G^{R,A}(t,t') = \mp i\theta(\pm(t-t'))\langle [a(t), a^\dagger(t')]_+\rangle$$

(iii) Check that the general relations:

$$G^R = G^{++} - G^{+-} = G^{-+} - G^{--}$$

$$G^A = G^{++} - G^{-+} = G^{+-} - G^{--}$$

between $G^{R,A}$ and $G^{\alpha,\beta}$ are verified.

(iv) Find the frequency representation for the retarded and advanced GFs ($G^{R,A}(\omega)$) and for the Keldysh GFs $G^{+-}(\omega)$ and $G^{-+}(\omega)$.

(v) From the general relations in (iii) obtain $G^{++}(\omega)$ and $G^{--}(\omega)$ for this model.

(vi) Verify the relations

$$G^{+-}(\omega) - G^{-+}(\omega) = G^A(\omega) - G^R(\omega)$$

$$G^{+-}(\omega) + G^{-+}(\omega) = (2n_F - 1)\left(G^A(\omega) - G^R(\omega)\right)$$

where $n_F = 1/(e^{\beta(\epsilon-\mu)} + 1)$. Which of these relations hold in general conditions?

Repeat the exercise for the bosonic case.

As mentioned above, knowledge of the Keldysh GFs allows us to obtain the mean value of any single-particle observable. For instance, it is straightforward to express the particle density and current as

$$n(r, t) = \mp i G^{+-}(x, x)$$

$$\mathbf{J}(r, t) = \frac{e\hbar}{2m} \left(\nabla_r - \nabla_{r'}\right) G^{+-}(x, x')|_{x=x'} . \tag{8.12}$$

It is thus convenient to directly focus on the perturbative expansion of $G(x, x')$, which is formally given by

$$G(x, x') = \sum_n \frac{(-i)^{n+1}}{n!} \int_{\mathcal{C}} dt_1 \dots \int_{\mathcal{C}} dt_n \langle \mathbf{T}_{\mathcal{C}} V_I(t_1) \dots V_I(t_n) \Psi(x) \Psi^{\dagger}(x') \rangle . \tag{8.13}$$

8.2 Perturbative Expansion for a One-Body Perturbation

Let us first consider the simple case of a scalar field $V(t) = \int dr \, \Psi^{\dagger}(x)\phi(r, t)\Psi(x)$. The $n = 0$ order is naturally given by $G^{(0)}(x, x')$ which we can represent diagrammatically by an arrow connecting x and x' (see Fig. 8.1). Then, to first order we have

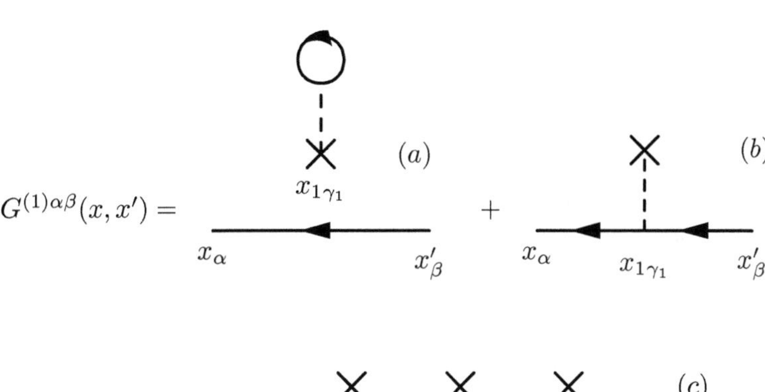

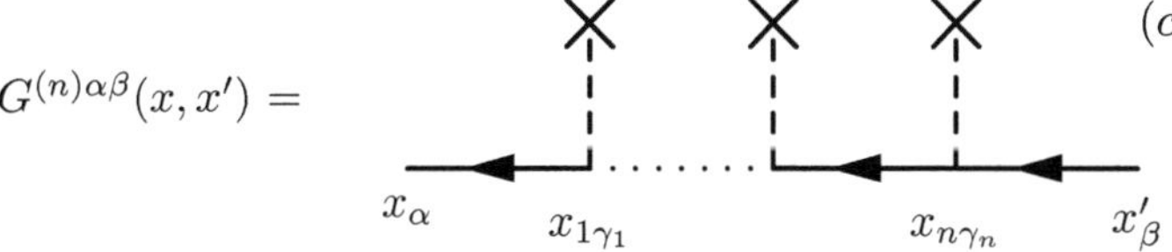

Fig. 8.1 Diagrams associated with a external potential perturbation. **a** and **b** correspond to the first order correction $G^{(1)}(x, x')$. The disconnected part (**a**) cancels when summed over the internal Keldysh index γ_1. **c** Generic order n correction. **d** Diagrammatic representation of the Keldysh–Dyson Equation (8.20) for the case of a external potential perturbation

$$G^{(1)}(x, x') = \frac{(-i)^2}{1!} \int_{\mathcal{C}} dt_1 \int dr_1 \langle \mathbf{T}_{\mathcal{C}} \Psi^\dagger(x_1)\Psi(x_1)\Psi(x)\Psi^\dagger(x') \rangle \,, \qquad (8.14)$$

and by applying Wick's theorem

$$\langle \mathbf{T}_{\mathcal{C}} \Psi^\dagger(x_1)\Psi(x_1)\Psi(x)\Psi^\dagger(x') \rangle = \langle \Psi^\dagger(x_1)\Psi(x_1) \rangle \langle \mathbf{T}_{\mathcal{C}} \Psi(x)\Psi^\dagger(x') \rangle$$
$$+ \langle \mathbf{T}_{\mathcal{C}} \Psi(x)\Psi^\dagger(x_1) \rangle \langle \mathbf{T}_{\mathcal{C}} \Psi(x_1)\Psi^\dagger(x') \rangle \,. \qquad (8.15)$$

The first term gives rise to *disconnected* diagrams, as shown in Fig. 8.1a. It is easy to see that within the non-equilibrium formalism these diagrams cancel when summed on the two branches of $\mathcal{C}$. Indeed,

$$G^{(1)A}(x, x') = \frac{(-i)^2}{1} \int_{\mathcal{C}} dt_1 \int dr_1 \phi(r_1, t_1) \langle \Psi^\dagger(x_1)\Psi(x_1) \rangle \langle \mathbf{T}_{\mathcal{C}} \Psi(x)\Psi^\dagger(x') \rangle$$
$$= \frac{(-i)^2}{1} \left[\int_{-\infty}^{\infty} dt_{1+} - \int_{-\infty}^{\infty} dt_{1-} \right] \phi(r_1, t_1) \langle \Psi^\dagger(x_1)\Psi(x_1) \rangle \langle \mathbf{T}_{\mathcal{C}} \Psi(x)\Psi^\dagger(x') \rangle \,,$$

which naturally vanishes as $\phi(r, t_{1+}) = \phi(r_1, t_{1-})$ for a physical field. We are thus left with

$$G^{(1)}(x, x') = G^{(1)B}(x, x') = -\int_{\mathcal{C}} dx_1 \langle \mathbf{T}_{\mathcal{C}} \Psi(x)\Psi^\dagger(x_1) \rangle \phi(x_1) \langle \mathbf{T}_{\mathcal{C}} \Psi(x_1)\Psi^\dagger(x') \rangle$$
$$= \int_{\mathcal{C}} dx_1 G^{(0)}(x, x_1) \phi(x_1) G^{(0)}(x_1, x') \,. \qquad (8.16)$$

Instead of dealing with integrations over $\mathcal{C}$, we can translate them into the more familiar single integrations from $-\infty$ to $+\infty$ by taking the explicit Keldysh components in (8.16), i.e.,

$$G^{(1)\alpha\beta}(x, x') = \int_{\mathcal{C}_+} dt_1 \int dr_1 G^{(0)\alpha+}(x, x_1) \phi(x_1) G^{(0)+\beta}(x_1, x')$$
$$+ \int_{\mathcal{C}_-} dt_1 \int dr_1 G^{(0)\alpha-}(x, x_1) \phi(x_1) G^{(0)-\beta}(x_1, x') \,, \qquad (8.17)$$

which can be cast into a 2×2 matrix form

$$\hat{G}^{(1)}(x, x') = \int_{-\infty}^{\infty} dt_1 \int dr_1 \hat{G}^{(0)}(x, x_1) \hat{\phi}(x_1) \hat{G}^0(x_1, x') \,, \qquad (8.18)$$

where we have defined

$$\hat{G}^{(0)} = \begin{pmatrix} G^{(0)++} & G^{(0)+-} \\ G^{(0)-+} & G^{(0)--} \end{pmatrix}, \quad \hat{\phi} = \begin{pmatrix} \phi & 0 \\ 0 & -\phi \end{pmatrix} .$$

It is then simple to convince ourselves that the n-order term in this case would be given by

$$\hat{G}^{(n)}(x, x') = \int dx_1 \ldots \int dx_n \hat{G}^{(0)}(x, x_1)\hat{\phi}(x_1) \ldots \hat{\phi}(x_n)\hat{G}^{(0)}(x_n, x') , \quad (8.19)$$

where now $\int dx \equiv \int_{-\infty}^{\infty} dt \int dr$. This correction is represented in Fig. 8.1c. The infinite series can thus simply expressed as

$$\hat{G}(x, x') = \hat{G}^{(0)}(x, x') + \int dx_1 \hat{G}^{(0)}(x, x_1)\hat{\phi}(x_1)\hat{G}(x_1, x') , \quad (8.20)$$

which corresponds to the *Dyson* equation for this elementary problem. This is diagrammatically represented in Fig. 8.1d, where the only difference with the equilibrium case is summation over branch indexes and taking into account the sign change of $\phi(x)$ on the $-$ branch.

8.3 Case of Many-Body Interactions

Let us consider a generic two-body interaction of the form

$$V = \frac{1}{2} \int dr_1 \int dr_2 \Psi^{\dagger}(x_1)\Psi^{\dagger}(x_2)v(x_1 - x_2)\Psi(x_2)\Psi(x_1) \quad (8.21)$$

where, for simplicity, we do not take into account the spin indexes. As is the case of direct or unscreened interactions $v(x_1 - x_2) = v(r_1 - r_2)$, i.e., V is an *instantaneous* interaction.

As in the simple example considered before, the diagrammatic rules for this case are the same as in the equilibrium case except for the additional branch index on each vertex which have to be summed up. Diagrams for the expansion of the single-particle GF up to second order are shown in Fig. 8.2. It should be noticed that only connected diagrams are included as the disconnected ones cancel in the Keldysh formalism for the same reasons as discussed in the case of the interaction with a one-body potential.

As in the equilibrium case, the single-particle self-energy can be defined as the sum of all irreducible diagrams, as illustrated in Fig. 8.2. Within the Keldysh formalism, however, the self-energy $\hat{\Sigma}(x, x')$, adopts a matrix form

$$\hat{\Sigma}(x, x') = \begin{pmatrix} \Sigma^{++}(x, x') & \Sigma^{+-}(x, x') \\ \Sigma^{-+}(x, x') & \Sigma^{--}(x, x') \end{pmatrix} . \quad (8.22)$$

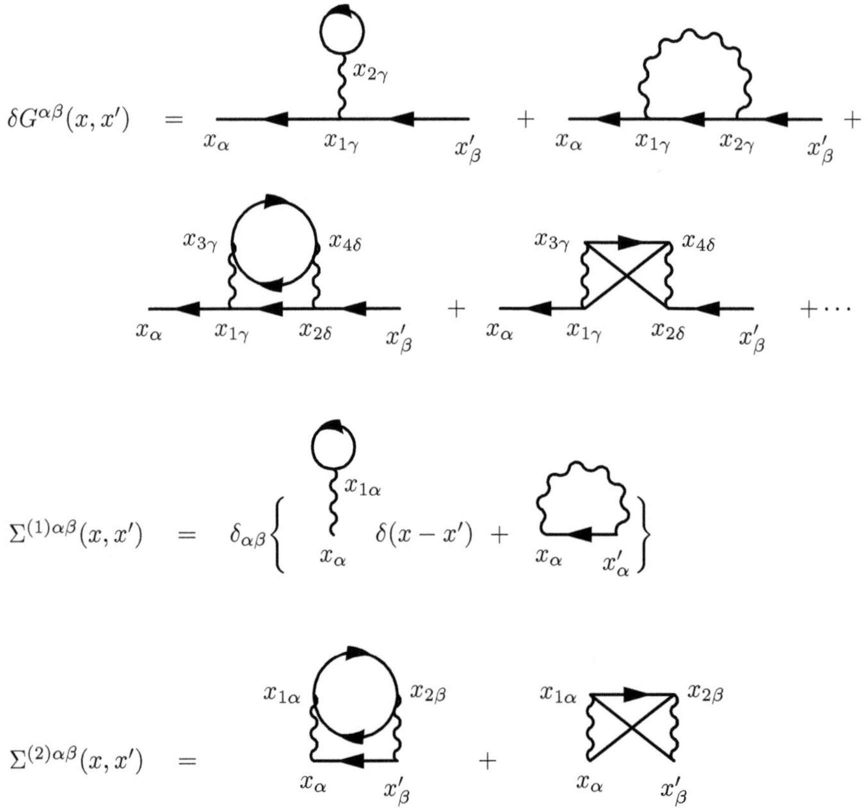

$$\delta G^{\alpha\beta}(x, x') \quad = \quad \cdots \quad + \quad \cdots \quad +$$

$$\cdots \quad + \quad \cdots \quad + \cdots$$

$$\Sigma^{(1)\alpha\beta}(x, x') \quad = \quad \delta_{\alpha\beta}\left\{ \quad \delta(x - x') \;+\; \quad \right\}$$

$$\Sigma^{(2)\alpha\beta}(x, x') \quad = \quad \cdots \quad + \quad \cdots$$

Fig. 8.2 Diagrams associated with a two-body interaction

The knowledge of this quantity allows to obtain the fully dressed GF by solving the Dyson equation

$$\hat{G}(x, x') = \hat{G}^{(0)}(x, x') + \int dx_1 \int dx_2 \hat{G}^{(0)}(x, x_1)\hat{\Sigma}(x_1, x_2)\hat{G}(x_2, x') , \quad (8.23)$$

which allows to identify $\hat{\Sigma}(x, x')$ with $\delta(x - x')\phi(x)\sigma_z$ for the case of a one-body potential seen in the previous section. Let us notice that, unlike this case, a many-body interaction leads to finite $\Sigma^{+-,-+}$ components which require higher order irreducible diagrams as the second order ones in Fig. 8.2.

> **Exercise II.2: Diagrams in Keldysh formalism**
> (i) Draw diagrams and write the corresponding expressions for the first order terms in the expansion of the single particle Green functions due to a two body interaction of the form
>
> $$V = \frac{1}{2} \int dr_1 \int dr_2 \; \psi^\dagger(x_1)\psi^\dagger(x_2)v(r_1 - r_2)\psi(x_2)\psi(x_1)$$
>
> (ii) Check cancellation of the disconnected diagrams up to first order.
> (iii) Draw an example of a connected second order diagram and write the corresponding expression.

8.4 The Dyson Equation as an Intregro-Differential Equation

As can be noticed, the Dyson equation (8.23) is an integral equation which might be difficult to solve in general conditions. We can get more insight on its meaning by converting it into a integro-differential equation. For this purpose we will take its time derivative $\partial_t \hat{G}(x, x')$. In the case of the unperturbed Keldysh GFs we have

$$\begin{pmatrix} (i\partial_t - h(x)) & 0 \\ 0 & -(i\partial_t - h(x)) \end{pmatrix} \hat{G}^{(0)}(x, x') = \hat{\delta}(x - x') , \qquad (8.24)$$

where $h(x)$ defines the non-interacting Hamiltonian $H_0 = \int dx\, \Psi^\dagger(x)h(x)\Psi(x)$. By combining this expression with Eq. (8.23) we thus obtain

$$\begin{pmatrix} (i\partial_t - h(x)) & 0 \\ 0 & -(i\partial_t - h(x)) \end{pmatrix} \hat{G}(x, x') = \hat{\delta}(x - x') + \int dx_1 \hat{\Sigma}(x, x_1)\hat{G}(x_1, x') ,$$

$$(8.25)$$

which can be viewed as a *quantum kinetic equation*, equivalent to the Boltzmann equation in the classical case.

8.5 The Triangular Representation

As mentioned in the previous section, not all four Keldysh GFs are independent as they are linked by Eq. (8.9). This feature allows to eliminate one of the GFs from the formalism, which can be accomplished in an elegant way by a simple $45°$ rotation in the $+, -$ space, as originally proposed by Keldysh (1964). One can take, for instance $\hat{U} = (1 - i\sigma_y)/\sqrt{2}$, so that

$$\hat{\tilde{G}} = \hat{U}\hat{G}\hat{U}^\dagger = \frac{1}{2}\begin{pmatrix} G^{++} - G^{-+} - \left(G^{+-} - G^{--}\right) & G^{++} - G^{-+} + \left(G^{+-} - G^{--}\right) \\ G^{++} + G^{-+} - \left(G^{+-} + G^{--}\right) & G^{++} + G^{-+} + \left(G^{+-} + G^{--}\right) \end{pmatrix} ,$$

$$\tag{8.26}$$

which, by virtue of (8.9) and the relation of Keldysh GFs with $G^{R,A}$ (8.11) leads to

$$\hat{\tilde{G}} = \begin{pmatrix} 0 & G^A \\ G^R & G^K \end{pmatrix} , \tag{8.27}$$

where we have defined $G^K = G^{+-} + G^{-+}$. Equation (8.27) defines what is called the *triangular* representation within the Keldysh formalism.

To visualize the full power of the approach it is necessary to perform the Keldysh rotation on the Dyson equation (8.23). For that purpose we can write Eq. (8.23) in a more compact way, as

$$\hat{G} = \hat{G}^{(0)} + \hat{G}^{(0)} \otimes \hat{\Sigma} \otimes \hat{G} , \tag{8.28}$$

where $\otimes$ indicates integration of time and coordinates internal arguments. In the rotated frame we would thus have

$$\hat{\tilde{G}} = \hat{\tilde{G}}^{(0)} + \hat{\tilde{G}}^{(0)} \otimes \hat{\tilde{\Sigma}} \otimes \hat{\tilde{G}} , \tag{8.29}$$

where $\hat{\tilde{A}}$ denotes $U\hat{A}U^\dagger$. Unlike $\hat{G}$, the self-energy matrix transforms through the Keldysh rotation as

$$\hat{\tilde{\Sigma}} = \begin{pmatrix} \Sigma^K & \Sigma^R \\ \Sigma^A & 0 \end{pmatrix} , \tag{8.30}$$

where

$$\begin{aligned} \Sigma^K &= -\left(\Sigma^{+-} + \Sigma^{-+}\right) \\ \Sigma^R &= \Sigma^{++} + \Sigma^{+-} = -\left(\Sigma^{--} + \Sigma^{-+}\right) \\ \Sigma^A &= \Sigma^{++} + \Sigma^{-+} = -\left(\Sigma^{--} + \Sigma^{+-}\right) . \end{aligned} \tag{8.31}$$

This difference is due to the fact that $\hat{\Sigma}$ has typically the structure $\hat{\gamma} \otimes \hat{G} \otimes \hat{\gamma}$, where $\hat{\gamma} \propto \sigma_z$ represents a vertex insertion with opposite signs on the two branches of $\mathcal{C}$. Finally, coming back to the rotated Dyson equation (8.29) one can easily show that the equations decouple into

$$G^{A,R} = G^{A,R(0)} + G^{A,R(0)}\Sigma^{A,R}G^{A,R} \tag{8.32}$$

$$G^K = G^{K(0)} + G^{R(0)}\Sigma^K G^A + G^{R(0)}\Sigma^R G^K + G^{K(0)}\Sigma^A G^A , \tag{8.33}$$

where integration over internal arguments is now implicitly assumed.

This decoupling suggests that, in general, non-equilibrium problems can be solved in two steps. First solve for $G^{R,A}$ and then introduce these quantities into the equation for the Keldysh component G^K. Notice that $G^K = G^{+-} + G^{-+}$ would allow to express mean values of physical observables, as already shown for G^{+-}.

It is also worth noticing that a more symmetrical expression of the equation for G^K can be obtained by taking into account Eq. (8.32). In fact, we can write (8.33) as

$$\left[I - G^{R(0)} \Sigma^R \right] G^K = G^{K(0)} + G^{R(0)} \Sigma^K G^A + G^{K(0)} \Sigma^A G^A , \qquad (8.34)$$

and using the identity $\left[I - G^{R(0)} \Sigma^R \right]^{-1} \equiv \left(I + G^R \Sigma^R \right)$, and Eq. (8.32)

$$G^K = \left(I + G^R \Sigma^R \right) G^{K(0)} \left(I + \Sigma^A G^A \right) + G^R \Sigma^K G^A . \qquad (8.35)$$

Exercise II.3: Triangular representation of Dyson equation
The Keldysh rotation $\hat{U} = (\hat{1} - i\hat{\sigma}_y)/\sqrt{2}$ transforms the single particle Keldysh Green functions into it's triangular form, i.e.,

$$\hat{U} \hat{G} \hat{U}^\dagger = \begin{pmatrix} 0 & G^A \\ G^R & G^K \end{pmatrix}$$

where $G^K = G^{+-} + G^{-+}$.
Starting from the Dyson equation in Keldysh space

$$\hat{G} = \hat{G}^{(0)} + \hat{G}^{(0)} \otimes \hat{\Sigma} \otimes \hat{G}$$

show that the necessary condition on $\hat{\Sigma}$ to keep the triangular form is that $\Sigma^{++} + \Sigma^{--} = -\left(\Sigma^{+-} + \Sigma^{-+} \right)$.
Show also that under this condition

$$\hat{U} \hat{\Sigma} \hat{U}^\dagger = \begin{pmatrix} \Sigma^K & \Sigma^R \\ \Sigma^A & 0 \end{pmatrix}.$$

8.6 Langreth Rules

We've seen that the perturbative expansion in Keldysh space has a 2×2 matrix structure, which can be then transformed into the A, R, K space by a simple rotation. There exists another useful trick to convert convolution of GFs over the $\mathcal{C}$ contour into $+-$, A, R components, known as *Langreth rules*. Let's first notice that one can write equations like (8.23) in a scalar way using contour integration, i.e.,

$$G(x, x') = G^{(0)}(x, x') + \int_{\mathcal{C}} dx_1 \int_{\mathcal{C}} dx_2 G^{(0)}(x, x_1)\Sigma(x_1, x_2)G(x_2, x') , \quad (8.36)$$

where $\int_{\mathcal{C}} dx \equiv \int_{\mathcal{C}} dt \int dr$ and summation over any other degree of freedom which might be involved. Thus, in general, we would need to evaluate convolutions of the type

$$A \otimes B = \int_{\mathcal{C}} A(t, t_1)B(t_1, t') , \quad (8.37)$$

from which we would like to know how to extract directly a $+-$, R or A component. Let us first consider how to obtain $(A \otimes B)^{+-}$

$$(A \otimes B)^{+-} = \int_{-\infty}^{\infty} dt_1 A^{++}(t, t_1)B^{+-}(t_1, t') + \int_{-\infty}^{\infty} dt_1 A^{+-}(t, t_1)B^{--}(t_1, t')$$

$$= \int_{-\infty}^{\infty} dt_1 \left[A^{++}(t, t_1)B^{+-}(t_1, t') - A^{+-}(t, t_1)B^{--}(t_1, t') \right] . \quad (8.38)$$

Then, taking into account that $A^{++} = A^R + A^{+-}$ and $B^{--} = B^{+-} - B^A$ we obtain

$$(A \otimes B)^{+-} = \int_{-\infty}^{\infty} dt_1 \left[A^R(t, t_1) + A^{+-}(t, t_1) \right] B^{+-}(t_1, t') - A^{+-}(t, t_1) \left[B^{+-}(t_1, t') - B^A(t_1, t') \right]$$

$$= \int_{-\infty}^{\infty} dt_1 \left[A^R(t, t_1)B^{+-}(t_1, t') + A^{+-}(t, t_1)B^A(t_1, t') \right] , \quad (8.39)$$

or, in a more compact way $(A \otimes B)^{+-} = A^R \otimes B^{+-} + A^{+-} \otimes B^A$. In a similar way (see exercise) one can demonstrate $(A \otimes B)^{A,R} = A^{A,R} \otimes B^{A,R}$.

Exercise II.4: Langreth rules

In the lectures we showed that the so-called Langreth rule to get a $+-$ component of a convolution of two functions on the Keldysh contour was given by

$$(A \otimes B)^{+-} = A^R \otimes B^{+-} + A^{+-} \otimes B^A \quad (8.40)$$

Show that the following relations hold for the $-+$ and the retarded, advanced components

$$(A \otimes B)^{-+} = A^R \otimes B^{-+} + A^{-+} \otimes B^A$$
$$(A \otimes B)^{R,A} = A^{R,A} \otimes B^{R,A} .$$

> **Exercise II.5: Dyson equation for G^{+-}**
> Starting from the Dyson equation defined on the Keldysh contour use the Langreth rules to show that
>
> $$G^{+-} = \left(1 + G^R \Sigma^R\right) g^{+-} \left(1 + \Sigma^A G^A\right) + G^R \Sigma^{+-} G^A \ . \tag{8.41}$$
>
> and a similar equation for G^{-+}.

We should notice that the self-energies Σ^{+-} and Σ^{-+} that are defined using Langreth rules differ in a sign with respect to the ones that are obtained using the full-Keldysh formalism diagrammatic rules. This is due to the extra $(-)$ sign for vertices on the $\mathcal{C}_-$ branch which is required to maintain the matrix structure in the full formalism.

8.7 Frequency Representation

Whenever we are dealing with a steady state or time-translational invariant situation GFs have the property $\hat{G}(t, t') = \hat{G}(t - t')$. One can thus get rid of time convolutions by working in the frequency representation, i.e., express quantities in terms of

$$\hat{G}(\omega) = \int_{-\infty}^{\infty} dt\, e^{i\omega t} \hat{G}(t, 0) \ , \tag{8.42}$$

which would satisfy a Dyson equation of the form

$$\hat{G}(\omega) = \hat{G}^{(0)}(\omega) + \hat{G}^{(0)}(\omega) \otimes \hat{\Sigma}(\omega) \otimes \hat{G}(\omega) \ . \tag{8.43}$$

Notice that this equation might still be hard to solve as it includes integration over spatial, spin or other internal degrees of freedom.

One can also write Eq. (8.25) in frequency representation as

$$\begin{pmatrix} (\omega - h(r)) & 0 \\ 0 & -(\omega - h(r)) \end{pmatrix} \hat{G}(r, r', \omega) = \hat{\delta}(r - r') + \int dr_1 \hat{\Sigma}(r, r_1, \omega) \hat{G}(r_1, r', \omega) \ , \tag{8.44}$$

which, after performing the Keldysh rotation decouples into

$$(\omega - h(r))\, G^{R,A}(r, r', \omega) = \delta(r - r') + \int dr_1 \Sigma^{R,A}(r, r_1, \omega) G^{R,A}(r_1, r', \omega)$$

$$(\omega - h(r))\, G^K(r, r', \omega) = \int dr_1 \Sigma^K(r, r_1, \omega) G^A(r_1, r', \omega) + \Sigma^R(r, r_1, \omega) G^K(r_1, r', \omega) \ .$$

$$\tag{8.45}$$

Chapter 9
Applications: Electron Transport at the Nanoscale

One of the more successful applications of the non-equilibrium GFs formalism has been the description of quantum transport in meso and nanoscale devices. At these scales and at low temperature, quantum coherence cannot be neglected which, combined with interactions, leads to peculiar and rather complex phenomena.

In a generic transport geometry a nanoscale region is connected to the *macroscopic* world through a series of contacts which allow to inject currents and measure voltages. A usual simplifying assumption in the description of quantum transport is that well inside the contacts thermal equilibrium is restored so that each contact α is characterized by a certain chemical potential μ_α and temperature T_α. On the other hand, at the nanoscale region a non-equilibrium distribution would be established whose determination is the main objective of the theory. One is interested in determining measurable quantities like the mean current on each contact, $I_\alpha(t) = \langle \hat{I}_\alpha(t) \rangle$ or current correlation functions of the type $\langle \delta \hat{I}_\alpha(t) \delta \hat{I}_\beta(t') \rangle$, where $\delta \hat{I}_\alpha = \hat{I}_\alpha - \langle \hat{I}_\alpha \rangle$. It should be noticed that in the quantum transport regime current fluctuations are significant compared to mean values. In addition, these properties are strongly influenced by electronic correlations like, for instance, in the presence of superconductivity of ferromagnetism in the leads.

As a first simple example let us consider the case of a normal, single channel tunnel junction as schematically represented in Fig. 9.1. This situation could be experimentally realized in the case of a metallic STM tip which approaches a metallic substrate, a modeled by a Hamiltonian in a local orbital basis $\hat{H} = \hat{H}_T + \hat{H}_S + \hat{V}$ where

$$\hat{H}_{T,S} = \sum_{i \in T,S} \varepsilon_i c_i^\dagger c_i + \sum_{\langle i,j \rangle} V_{ij} c_i^\dagger c_j + \text{h.c.} - \mu_{T,S} \hat{N}_{T,S}$$

$$\hat{V} = \sum_{i \in T, j \in S} V' c_i^\dagger c_j + \text{h.c.} \tag{9.1}$$

where $c_i^\dagger$ (c_i) creates (destroys) an electron at a site i which can be either located at the tip (T) or sample (S) region. In this first example we just consider the case of spinless electrons.

© The Author(s), under exclusive license to Springer Nature Switzerland AG 2024   115
J. Merino and A. L. Yeyati, *Many-Body Techniques in Condensed Matter Physics*,
UNITEXT for Physics, https://doi.org/10.1007/978-3-031-55143-7_9

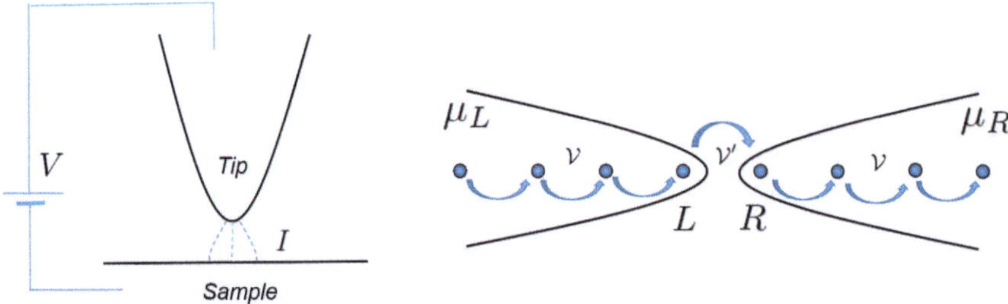

Fig. 9.1 Left panel: scheme of a typical nanoscale transport experiment using an STM. Right panel: single channel model for a normal contact

In order to determine the expression for the current operator within this basis let us analyze the equation of motion for the charge operator $-e\hat{n}_i = -ec_i^\dagger c_i$

$$-e\partial_t \hat{n}_i = -\frac{ie}{\hbar}\left[\hat{H}, \hat{n}_i\right] = -\frac{ie}{\hbar}\sum_{\langle i \neq j\rangle}\left(\mathcal{V}_{ij}c_j^\dagger c_i - \mathcal{V}_{ji}c_i^\dagger c_j\right) \equiv \sum_{\langle i \neq j\rangle}\hat{I}_{ij} \,, \qquad (9.2)$$

which allows us to identify $\hat{I}_{ij} = ie/\hbar(\mathcal{V}_{ij}c_i^\dagger c_j - \text{h.c.})$.

It is thus straightforward to write an expression for the mean current in terms of Keldysh GFs

$$\langle \hat{I}_{ij}\rangle(t) = \frac{e}{\hbar}\left(\mathcal{V}_{ij}G_{ji}^{+-}(t,t) - \mathcal{V}_{ji}G_{ij}^{+-}(t,t)\right) \,. \qquad (9.3)$$

We can further simplify the model into a one-dimensional problem where there is a weak link with a hopping term $\mathcal{V}' \in \mathbb{R}$ coupling the left (L) and right (R) regions as illustrated in Fig. 9.1 (right panel). In that case (9.3) becomes

$$\langle \hat{I}_{LR}\rangle = \frac{e\mathcal{V}'}{\hbar}\left(G_{RL}^{+-}(t,t) - G_{LR}^{+-}(t,t)\right) = \frac{e\mathcal{V}'}{\hbar}\int\frac{d\omega}{2\pi}\left(G_{RL}^{+-}(\omega) - G_{LR}^{+-}(\omega)\right) \,. \qquad (9.4)$$

Let us now discuss how to obtain the GFs entering in Eq. (9.4). We can decompose $H = H_0 + V$, where $V \to 0$ for $t \to -\infty$ contains the left-right coupling $\mathcal{V}'$. In this way the system in the remote past is in a local thermal equilibrium with chemical potentials $\mu_{L,R}$, such that $eV = \mu_L - \mu_R$. In order to obtain G_{LR}^{+-} and G_{RL}^{+-} we must solve the corresponding Dyson equations where the self-energy $\hat{\Sigma}$ describes the tunneling processes between left and right. It is easy to see that in Keldysh space we have

$$\hat{\Sigma} = \mathcal{V}'\sigma_z\left(|L\rangle\langle R| + |R\rangle\langle L|\right) \,, \qquad (9.5)$$

which can be written in a more compact way as $\check{\hat{\Sigma}} = \mathcal{V}'\sigma_z\tau_x$, where the $\check{\ }$ symbol denotes $L - R$ space and τ_x is a Pauli matrix in that space. Performing the Keldysh rotation we get $\check{\Sigma}^{R,A} = \mathcal{V}'\sigma_x$ and $\check{\Sigma}^K = \check{\Sigma}^{+-} = 0$. Thus, the Dyson equation $\check{G}^{+-}$ reads

$$\check{G}^{+-} = \left(\check{I} + \check{G}^R \check{\Sigma}^R\right) \check{g}^{+-} \left(\check{I} + \check{\Sigma}^A \check{G}^A\right) , \tag{9.6}$$

where $\check{g} = \mathrm{diag}(g_L, g_R)$ stands for the uncoupled leads GFs. In the same way, for $\check{G}^{R,A}$ we have

$$\check{G}^{R,A} = \check{g}^{R,A} \left(\check{I} + \check{\Sigma}^{R,A} \check{G}^{R,A}\right) . \tag{9.7}$$

To solve the problem we thus need a model for $\check{g}^{A,R,+-}$. Let's start by stating some general properties of retarded-advanced GFs within local orbital models. For a non-interacting problem we formally have

$$\mathcal{G}^{R,A}(\omega) = \lim_{\eta \to 0} \left[\omega \pm i\eta - h(r)\right]^{-1} , \tag{9.8}$$

where the infinitesimal quantity η allows to select the appropriate integration contour for the retarded-advanced component when transforming into the time-representation. One can further use a eigenstate basis $\{|\Phi_n\rangle\}$ to get

$$\mathcal{G}_{ij}^{R,A}(\omega) = \lim_{\eta \to 0} \sum_n \frac{\langle i|\Phi_n\rangle\langle\Phi_n|j\rangle}{\omega - E_n \pm i\eta} . \tag{9.9}$$

From this expression it is straightforward to see the relation with the local density of states (LDOS) $\rho_i(\omega)$

$$\mathcal{G}_{ii}^{A}(\omega) - \mathcal{G}_{ii}^{R}(\omega) = 2\pi i \sum_n |\langle i|\Phi_n\rangle|^2 \delta(\omega - E_n) \equiv 2\pi i \rho_i(\omega) . \tag{9.10}$$

One-dimensional (1D) models provide a simple playground to obtain analytical expressions for the required unperturbed GFs. Let us first consider the case of an infinite homogeneous 1D chain, with eigenvalues $\varepsilon(k) = 2\mathcal{V}\cos(k)$ for which Eq. (9.9) yields

$$\mathcal{G}_{ij}^{R,A}(\omega) = \frac{1}{2\pi} \int_{-\pi}^{\pi} dk \frac{e^{i(i-j)k}}{\omega \pm i0^+ - \varepsilon(k)} , \tag{9.11}$$

where 0^+ now accounts for the imaginary infinitesimum η. This integral can be computed by transforming it into a complex variable integral, by defining $z = e^{ik}$. Then

$$\mathcal{G}_{ij}^{R,A}(\omega) = \frac{-1}{2\pi i \mathcal{V}} \oint_{|z|=1} dz \frac{z^{|i-j|}}{z^2 - 2zx_\pm + 1} , \tag{9.12}$$

where $x_\pm = (\omega \pm i0^+)/2\mathcal{V}$. The integrand have poles at $z_{1,2} = x \mp \sqrt{x^2 - 1}$, so that only the one such that $|z| < 1$ contributes to the integral. The proper branch of the square root for each GF can be taken either using a mathematical procedure (using the $\eta \to 0$ limit) or by physical considerations. Let us use this second procedure:

when $|x| \leq 1$ we are inside the band so that $z_{1,2} = x \pm i\sqrt{1 - x^2} = e^{\pm i\phi}$, where $\phi = \arccos(x)$. We thus have

$$\mathcal{G}_{ij}^{R,A}(\omega) = \frac{\mp i e^{\mp i|i-j|\phi}}{2\mathcal{V}\sqrt{1 - x^2}} \,, \qquad |\omega| < 2\mathcal{V}. \tag{9.13}$$

In contrast, for $|x| > 1$, i.e., outside the band, $z_{1,2} \in \mathbb{R}$ and we have

$$\mathcal{G}_{ij}^{R,A}(\omega) = \mathrm{sgn}(x)\frac{z_1^{|i-j|}}{2\mathcal{V}\sqrt{x^2 - 1}} \,, \qquad |\omega| > 2\mathcal{V}. \tag{9.14}$$

In Fig. 9.2a we show the behavior of the local $\mathcal{G}_{ii}^{R,A}(\omega)$ components, which exhibit the typical 1D singularities at the band edges $\omega = \pm 2\mathcal{V}$. From this 1D chain model one can derive the GFs corresponding to a semi-infinite wire, which could represent the uncoupled leads in Eqs. (9.6) and (9.7). For that purpose we introduce a perturbation $\hat{V} = \epsilon_0 |0\rangle\langle 0|$ at site 0 in the chain so that, when $\epsilon_0 \to \infty$ the chain is broken into two semi-infinite pieces. We thus get (see Exercise **II.6**)

$$g^{R,A}(\omega) = \mathcal{G}_{11}^{R,A} - \mathcal{G}_{10}^{R,A}\left(\mathcal{G}_{00}^{R,A}\right)^{-1}\mathcal{G}_{01}^{R,A} \,, \tag{9.15}$$

which leads to

$$g^{R,A}(\omega) = \frac{1}{\mathcal{V}}\left(x \mp i\sqrt{1 - x^2}\right) \,, \qquad |x| \leq 1 \,, \tag{9.16}$$

whereas outside the band the proper choice for the signs in the square root is

$$g^{R,A}(\omega) = \frac{1}{\mathcal{V}}\left(x - \mathrm{sgn}(\omega)\sqrt{x^2 - 1}\right) \,, \qquad |x| > 1 \,. \tag{9.17}$$

The real and imaginary parts of $g^{R,A}$ are shown in Fig. 9.2b.

> **Exercise II.6: Uncoupled leads GFs**
>
> Demonstrate that a divergent site level $\epsilon_0 \to \infty$ introduced to a infinite linear chain leads to Eq. (9.15) for the boundary GFs, $g^{R,A}$, on a semi-infinite chain. Alternatively, demonstrate that $g^{R,A}$ should satisfy the equation
>
> $$g^{R,A}(\omega) = \left[\omega - \mathcal{V}^2 g^{R,A}(\omega)\right]^{-1} \,,$$
>
> from which Eqs. (9.16), (9.17) can be directly obtained.

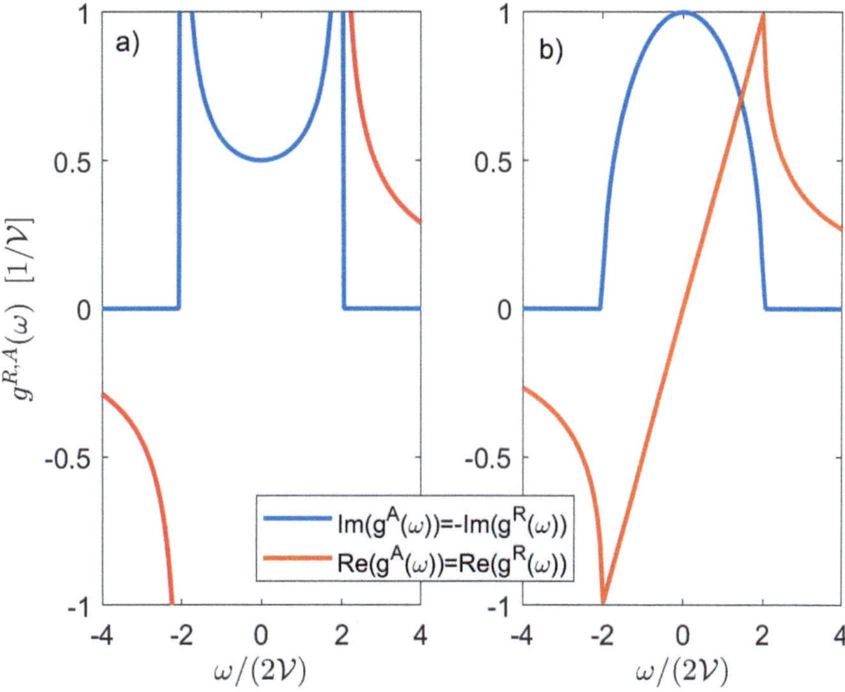

Fig. 9.2 Boundary GFs for single channel model: starting from the results for translational invariant 1D model (**a**) one obtains the boundary GFs for a semi-infinite chain (**b**) using Eq. (9.15)

As a next step we need to define $g_{L,R}^{+-}$. Due to the fact that the uncoupled leads can be considered to be in local thermal equilibrium we have

$$g_{L,R}^{+-}(\omega) = n_{L,R}(\omega)\left(g_{L,R}^{A}(\omega) - g_{L,R}^{R}(\omega)\right) , \tag{9.18}$$

where $n_{L,R}(\omega) = 1/(e^{\beta_{L,R}(\omega - \mu_{L,R})} + 1)$ are the Fermi factors corresponding to each lead. Notice that, in principle, not only the chemical potentials but also the temperatures, $T_{L,R} \equiv 1/\beta_{L,R}$, can be different. On the other hand $g_{L,R}^{A}(\omega) - g_{L,R}^{R}(\omega) = 2\pi i \rho_{L,R}(\omega)$, where $\rho_{L,R}$ is the LDOS at each lead.

Once we have defined the uncoupled leads GFs we can get to Eq. (9.6) to obtain the components G_{LR}^{+-} and G_{RL}^{+-} needed for the current. We have

$$G_{LR}^{+-} = \sum_{\mu} \left(\tilde{I} + \tilde{G}^{R}\tilde{\Sigma}^{R}\right)_{L\mu} g_{\mu}^{+-} \left(\tilde{I} + \tilde{\Sigma}^{A}\tilde{G}^{A}\right)_{\mu R}$$

$$= \left(1 + G_{LR}^{R}\mathcal{V}'\right) g_{L}^{+-}\mathcal{V}'G_{RR}^{A} + G_{LL}^{R}\mathcal{V}'g_{R}^{+-}\left(1 + \mathcal{V}'G_{LR}^{A}\right) , \tag{9.19}$$

which can be simplified from Eq. (9.7), using

$$G^A_{RR} = g^A_R \left(1 + V'G^A_{LR}\right) \text{ and } G^R_{LL} = \left(1 + G^R_{LR}V'\right) g^R_L \,,$$

leading to

$$G^{+-}_{LR} = |1 + V'G^A_{LR}|^2 \left(g^{+-}V'g^A_R + g^R_L V'g^{+-}_R\right) \,, \tag{9.20}$$

and in terms of the unperturbed GFs

$$G^{+-}_{LR} = \frac{\left(g^{+-}V'g^A_R + g^R_L V'g^{+-}_R\right)}{|1 - V'^2 g^A_L g^A_R|^2} \,. \tag{9.21}$$

It is then straightforward to express the current in the usual Landauer form

$$I_{LR}\frac{e}{h} \int d\omega \tau(\omega)\left(n_L(\omega) - n_R(\omega)\right) \,, \tag{9.22}$$

where, in terms of GFs, the transmission coefficient $\tau(\omega)$ can be written as (Ferrer et al. 1988)

$$\tau(\omega) = \frac{4\pi^2 V'^2 \rho_L(\omega)\rho_R(\omega)}{|1 - V'^2 g^A_L(\omega)g^A_R(\omega)|^2} \,. \tag{9.23}$$

It can be demonstrated that under general conditions $\tau(\omega)$ is bounded between 0 and 1 (Exercise **II.7**). Equation (9.23) adopts a particularly simple form in the low energy limit, $|\omega| << V$. In fact, in that limit we can approximate $g^A_{L,R} \simeq i/|V|$ and $\rho_{L,R} \simeq 1/(\pi|V|)$, so that

$$\tau(\omega) \simeq \frac{4\alpha}{(1 + \alpha)^2} \,, \tag{9.24}$$

where $\alpha = (V'/V)^2$, which is clearly bounded to $(0, 1)$.

Exercise II.7: Single channel contact transmission

Show that the transmission coefficient in this model given by the expression

$$\tau(\omega) = \frac{4V'^2 \text{Img}^A_L(\omega)\text{Img}^A_R(\omega)}{|1 - V'^2 g^A_L(\omega)g^A_R(\omega)|^2}$$

is bounded between 0 and 1 even when the frequency dependence of the leads GFs is taken into account.

9.1 Current Fluctuations

As mentioned in the introduction of this section, current fluctuations are significant in meso and nanoscale devices. They can be characterized by the current noise spectral density $S_I(\omega)$

$$S_I(\omega) = \int dt e^{i\omega t} \langle \delta\hat{I}(t)\delta\hat{I}(0) + \delta\hat{I}(0)\delta\hat{I}(t)\rangle \ . \tag{9.25}$$

While conductors at a macroscopic scale are characterized by *thermal* noise, quantum fluctuations lead to additional contributions to noise in micro and nanoscale devices at low temperatures. This is the case of the so-called shot noise, which is associated with the discrete and fermionic character of carriers in such devices.

In order to exemplify these phenomena let us go back to the simple 1D models based on local orbitals. The current operator between the leads, including spin can be written as

$$\hat{I}_{LR} = \frac{ie}{\hbar} \sum_\sigma \left(\mathcal{V}' c_{L\sigma}^\dagger c_{R\sigma} - \text{h.c.} \right) \ . \tag{9.26}$$

Introducing this expression into Eq. (9.25) and applying Wick theorem one can express $S_I(\omega)$ as

$$\begin{aligned}
S_I(\omega) = \frac{2e^2\mathcal{V}'^2}{\hbar^2} \int \frac{d\Omega}{2\pi} \Big[& G_{LL}^{+-}(\Omega - \omega)G_{RR}^{-+}(\Omega) + G_{RR}^{+-}(\Omega - \omega)G_{LL}^{-+}(\Omega) \\
& - G_{RL}^{+-}(\Omega - \omega)G_{RL}^{-+}(\Omega) - G_{LR}^{+-}(\Omega - \omega)G_{LR}^{-+}(\Omega) + (\omega \to -\omega) \Big] \ ,
\end{aligned} \tag{9.27}$$

where the factor 2 in front appears due to spin degeneracy. In order to evaluate this expression and get simple analytical results we can restrict ourselves to the low energy approximation for the uncoupled GFs, for which we have

$$G_{LR}^{+-}(\omega) = -\frac{\mathcal{V}'}{\mathcal{V}} \frac{(n_L(\omega) - n_R(\omega))}{(1+\alpha)^2}$$

$$G_{LL,RR}^{+-}(\omega) = \frac{2i}{\mathcal{V}} \frac{(n_{L,R}(\omega) + \alpha n_{R,L}(\omega))}{(1+\alpha)^2} \ , \tag{9.28}$$

and replacing $n_{L,R}$ by $(n_{L,R} - 1)$ in order to obtain $G_{\mu\nu}^{-+}$. Then, substituting into Eq. (9.27) at zero frequency, we obtain

$$\begin{aligned}
S_I(0) = \frac{4e^2}{h} \int d\Omega \Big\{ & \tau^2 \left[n_L(\Omega)(1 - n_L(\Omega)) + n_R(\Omega)(1 - n_R(\Omega)) \right] \\
& + \tau(1-\tau) \left[n_L(\Omega)(1 - n_R(\Omega)) + n_R(\Omega)(1 - n_L(\Omega)) \right] \Big\} \ ,
\end{aligned} \tag{9.29}$$

where τ denotes the junction transmission in the low-energy approximation, as given by Eq. (9.24). As shown in Büttiker (1990), Eq. (9.29) can also be derived using scattering theory.

Exercise II.8: Current noise in single channel contact

(i) Demonstrate that Eq. (9.29) can be written as

$$S(0) = \frac{4e^2}{h} \int d\omega' \left\{ \tau \left[n_L(\omega') \left(1 - n_R(\omega')\right) + n_R(\omega') \left(1 - n_L(\omega')\right) \right] \right.$$
$$\left. - \tau^2 \left[n_L(\omega') - n_R(\omega') \right]^2 \right\},$$

(ii) Analyze the different limits (i.e., zero temperature and zero bias) to see that the expected results are recovered.

9.2 Full Counting Statistics

Current fluctuations are not fully characterized by the noise spectrum. As any stochastic process, charge transport through a nanoscale device should have an associated probability distribution, the current noise being just related to the second moment of it. In this section we will show that the non-equilibrium formalism allows us to get access to the full distribution (what is call the "full counting statistics") by essentially the same effort than what is needed to calculate the mean current.

The aim of the theory is to characterize the probability distribution of the observable $\hat{N}(t_0) = \int_0^{t_0} dt\, \hat{I}(t)/e$ associated with the number of charges that traverse the nanoscale device (a junction or any other) in a given time interval t_0. Let us concentrate in our single channel model for which the current operator can be generated as a derivative of the tunnel Hamiltonian with respect to a *counting field* χ, i.e.:

$$V_\chi(t) = \sum_\sigma \mathcal{V}' \left(e^{i\chi(t)/2} c_{L\sigma}^\dagger(t) c_{R\sigma}(t) + \text{h.c.} \right)$$

$$I_\chi(t)/(2e) = \frac{\partial V_\chi}{\partial \chi} = \frac{i}{2} \sum_\sigma \mathcal{V}' \left(e^{i\chi(t)} c_{L\sigma}^\dagger(t) c_{R\sigma}(t) - \text{h.c.} \right) , \tag{9.30}$$

where $\chi(t) = \pm\chi$ change sign on the two branches of the closed contour $\mathcal{C}_0$ indicated in Fig. 9.3 With this definition the quantity $Z(\chi, t_0)$

$$Z(\chi, t_0) = \langle \mathcal{S}_{\mathcal{C}_0}^\chi \rangle = \left\langle \mathbf{T}_{\mathcal{C}_0} \exp\left(-i \int_{\mathcal{C}_0} dt\, V_\chi(t) \right) \right\rangle , \tag{9.31}$$

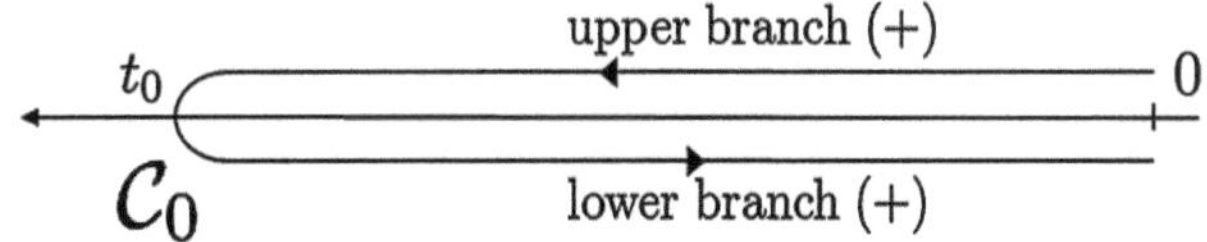

Fig. 9.3 Contour considered for the Full Couting Statistics (FCS) analysis

behaves as a generating function for $\hat{N}(t_0)$, i.e., it generates the moments $M_n(t_0)$ of its probability distribution as

$$M_n(t_0) = \langle \hat{N}^n(t_0) \rangle = (-i)^n \frac{\partial^n Z(\chi, t_0)}{\partial \chi^n} \big|_{\chi=0} . \tag{9.32}$$

One can also define $S(\chi, t_0) = \ln Z(\chi, t_0)$, which corresponds to a *cumulant generating function* (CGF), with the property

$$C_n(t_0) = \langle\langle \hat{N}^n(t_0) \rangle\rangle = (-i)^n \frac{\partial^n S}{\partial \chi^n} \big|_{\chi=0} , \tag{9.33}$$

where the cumulants C_n are related to the moments M_n by $C_1 = M_1, C_2 = M_2 - M_1^2$, $C_3 = M_3 - 3M_2 M_1 + 2M_1^3$, etc.

In order to make progress let's consider the first derivative of the generating function (9.31)

$$\frac{\partial Z(\chi, t_0)}{\partial \chi} = -i \int_{C_0} dt \left\langle \mathbf{T}_{C_0} \mathcal{S}_{C_0}^\chi \frac{\partial V_\chi}{\partial \chi}(t) \right\rangle . \tag{9.34}$$

By virtue of the linked cluster theorem we can decompose the integrand in (9.34) as

$$\langle \mathbf{T}_{C_0} \mathcal{S}_{C_0}^\chi \frac{\partial V_\chi}{\partial \chi}(t) \rangle = \left\langle \mathbf{T}_{C_0} \mathcal{S}_{C_0}^\chi \frac{\partial V_\chi}{\partial \chi}(t) \right\rangle_{conn} \langle \mathcal{S}_{C_0}^\chi \rangle , \tag{9.35}$$

where $\langle A \rangle_{conn}$ accounts to including connected diagrams only in its perturbative expansion. Notice that, in contrast to the $\chi = 0$ case, the factor $\langle \mathcal{S}_{C_0}^\chi \rangle = Z(\chi, t_0) \neq 1$. The expansion of this factor leads to disconnected diagrams which cancel when $\chi = 0$. From Eqs. (9.34) and (9.35) we thus obtain

$$\frac{\partial \ln Z(\chi, t_0)}{\partial \chi} = -i \int_{C_0} dt \left\langle \mathbf{T}_{C_0} \mathcal{S}_{C_0}^\chi \frac{\partial V_\chi}{\partial \chi}(t) \right\rangle_{conn} . \tag{9.36}$$

The next step would be to express Eq. (9.36) in terms of GFs. From (9.30) we have

$$i \frac{\partial \ln Z(\chi, t_0)}{\partial \chi} = \int_{\mathcal{C}_0} \frac{\mathcal{V}'}{2} \left(e^{i\chi(t)/2} G_{RL}(t,t) - e^{i\chi(t)/2} G_{LR}(t,t) \right)$$

$$= \frac{1}{2} \int_0^{t_0} \mathrm{Tr} \left[\sigma_z \left(\hat{\mathcal{V}}' \hat{G}_{RL}(t,t) - \hat{\mathcal{V}}'^\dagger \hat{G}_{LR}(t,t) \right) \right] , \tag{9.37}$$

where $\hat{\mathcal{V}}' = \mathcal{V}' \sigma_z e^{i\chi \sigma_z/2}$ and Tr indicates trace in Keldysh space. Notice that due to the presence of the counting field χ, breaking the symmetry between the two branches in the Keldysh contour, the relation $G^{++} + G^{--} = G^{+-} + G^{-+}$ is no longer valid and reduction to the triangular representation is thus not possible. Working with the full Keldysh representation is thus the "price to pay" in order to get access to the full counting statistics.

We can further simplify Eq. (9.37) using the Dyson equation in Keldysh space

$$\hat{G}_{RL} = \hat{g}_R \hat{\mathcal{V}}'^\dagger \hat{g}_L \left(\hat{I} - \hat{\mathcal{V}}' \hat{g}_R \hat{\mathcal{V}}'^\dagger \hat{g}_L \right)^{-1} , \tag{9.38}$$

and a similar expression for $\hat{G}_{LR}$. If we focus in the $t_0 \to \infty$ limit, which is relevant for actual experiments, stationary state conditions can be assumed and we can replace $\int_0^{t_0}$ by $t_0 \int d\omega/2\pi$. We thus get

$$i \frac{\partial \ln Z(\chi, t_0)}{\partial \chi} = \frac{t_0}{2} \int_{-\infty}^{\infty} \frac{d\omega}{2\pi} \mathrm{Tr} \left[\sigma_z \left(\hat{\mathcal{V}}' \hat{g}_R \hat{\mathcal{V}}'^\dagger \hat{g}_L \left(\hat{I} - \hat{\mathcal{V}}' \hat{g}_R \hat{\mathcal{V}}'^\dagger \hat{g}_L \right)^{-1} \right. \right.$$

$$\left. \left. - \left(\hat{I} - \hat{g}_L \hat{\mathcal{V}}' \hat{g}_R \hat{\mathcal{V}}'^\dagger \right)^{-1} \hat{g}_L \hat{\mathcal{V}}' \hat{g}_R \hat{\mathcal{V}}'^\dagger \right) \right] . \tag{9.39}$$

If we now define $\hat{A}(\chi) = \left(\hat{I} - \hat{\mathcal{V}}' \hat{g}_R \hat{\mathcal{V}}'^\dagger \hat{g}_L \right)$ we can use Jacobi's formula for a square matrix, i.e.:

$$\frac{\partial \ln \det \hat{A}}{\partial \chi} = \mathrm{Tr} \left[\hat{A}^{-1} \frac{\partial \hat{A}}{\partial \chi} \right] , \tag{9.40}$$

to show that

$$\frac{\partial \ln Z(\chi, t_0)}{\partial \chi} = t_0 \int_{-\infty}^{\infty} \frac{d\omega}{2\pi} \frac{\partial}{\partial \chi} \ln \det \left[\hat{I} - \hat{\mathcal{V}}' \hat{g}_R \hat{\mathcal{V}}'^\dagger \hat{g}_L \right] , \tag{9.41}$$

and finally the CGF can be written as

$$S(\chi, t_0) = t_0 \int_{-\infty}^{\infty} \frac{d\omega}{2\pi} \ln \det \left(\left[\hat{I} - \hat{\mathcal{V}}' \hat{g}_R \hat{\mathcal{V}}'^\dagger \hat{g}_L \right] \left[\hat{I} - \hat{\mathcal{V}}' \hat{g}_R \hat{\mathcal{V}}'^\dagger \hat{g}_L \right]_{\chi=0}^{-1} \right) ,$$

$$\tag{9.42}$$

Exercise II.9: Full Counting Statistics in the single channel model

(i) Use the low energy approximation for $\hat{g}_{L,R}$ to obtain from Eq. (9.42) the Levitov–Lesovik formula (Levitov and Lesovik 1993)

$$S(\chi, t_0) = \frac{t_0}{2\pi} \int d\omega \ln \left[1 + \tau \left(e^{i\chi} - 1\right) n_L(\omega) \left(1 - n_R(\omega)\right) \right.$$
$$\left. + \tau \left(e^{-i\chi} - 1\right) n_R(\omega) \left(1 - n_L(\omega)\right)\right]$$

(ii) Check that the expected results are obtained for the first two cumulants C_1 and C_2.

(iii) Find C_3 and discuss it's physical meaning.

9.3 Superconducting Transport

Superconducting nanostructures provide remarkable examples of phenomena which can be addressed using non-equilibrium many-body techniques. In the subsequent discussion we shall assume that the reader is familiar with the basic phenomenology associate to superconductivity (like the resistive transition, the Meissner effect or flux quantization) and start from the BCS pairing model describing a bulk conventional superconductor, which is given by

$$H_{BCS} = \sum_{k\sigma} \varepsilon_k c_{k\sigma}^\dagger c_{k\sigma} + \frac{1}{2} \sum_{kk',\sigma} V_{kk'} c_{k\sigma}^\dagger c_{-k\bar{\sigma}}^\dagger c_{-k'\bar{\sigma}} c_{k'\sigma} \tag{9.43}$$

with the simplified interaction

$$V_{kk'} \begin{cases} -V_0 & |\varepsilon_{k,k'} - \varepsilon_F| < \hbar\omega_D \\ 0 & \text{otherwise} \end{cases}$$

where the attractive interaction $-V_0$ is limited to states close to the Fermi surface with an energy distance fixed by the Debye frequency ω_D.

In the weak coupling limit, i.e., $V_0 \rho_F \ll 1$, where ρ_F is the Fermi level DOS, this model is well described by a superconducting mean field approach, which corresponds to replacing H_{BCS} by

$$H_{MF} = \sum_{k\sigma} (\varepsilon_k - \mu) c_{k\sigma}^\dagger c_{k\sigma} + \sum_k \Delta_k c_{k\uparrow}^\dagger c_{-k\downarrow}^\dagger + \text{h.c.} , \tag{9.44}$$

where, at zero temperature

$$\Delta_k = \sum_{k'} V_{kk'} \langle c_{-k'\downarrow} c_{k'\uparrow} \rangle = -\frac{1}{2} \sum_{k'} \frac{V_{kk'} \Delta_{k'}}{\sqrt{(\varepsilon_{k'} - \varepsilon_F)^2 + \Delta_{k'}^2}} , \tag{9.45}$$

whose self-consistent solution leads to $\Delta_k \simeq 2\hbar\omega_D \exp(-1/V_0\rho_F)$ for $|\varepsilon_k - \varepsilon_F| < \hbar\omega_D$ and $\Delta_k = 0$ otherwise.

For describing transport in nanoscale devices we are typically interested in non-homogeneous situations where the electrostatic potential and the pairing amplitude may vary in space. For that one can think of a continuous model of the form

$$H_{MF} = \int dr \sum_{\sigma} \psi_{\sigma}^{\dagger}(r) \left[-\frac{\hbar^2 \nabla^2}{2m} + V(r) - \mu \right] \psi_{\sigma}(r) + \left(\Delta(r)\psi_{\uparrow}^{\dagger}(r)\psi_{\downarrow}^{\dagger}(r) + \text{h.c.} \right) . \tag{9.46}$$

Notice that the local pairing interaction in Hamiltonian (9.46) is consistent with the momentum independent pairing in the BCS model (i.e., s-wave pairing). For transport calculations of the type we presented in the previous sections it is convenient to use a discretized version of this model, which can be written as

$$H_{MF} = \sum_{i\sigma} (\varepsilon_i - \mu) \, c_{i\sigma}^{\dagger} c_{i\sigma} + \sum_{\langle i,j\rangle,\sigma} \left(V_{ij} c_{i\sigma}^{\dagger} c_{j\sigma} + \text{h.c.} \right)$$
$$+ \sum_i \left(\Delta_i c_{i\uparrow}^{\dagger} c_{i\downarrow}^{\dagger} + \text{h.c.} \right) , \tag{9.47}$$

where Δ_i is a spatial dependent pairing amplitude, which typically would take a fixed value Δ in the superconducting regions and 0 in the normal ones.

9.3.1 Nambu Formalism

The broken symmetry associated with superconductivity is most conveniently described using the Nambu formalism, where instead of the usual fermionic operators quantities are expressed in terms of Nambu spinors $\hat{\psi}_i = \left(c_{i\uparrow} \; c_{i\downarrow}^{\dagger} \right)^T$. Thus, Hamiltonian (9.47) can be written as

$$H_{MF} = \sum_i \hat{\psi}_i^{\dagger} \hat{h}_i \hat{\psi}_i + \sum_{\langle i,j\rangle} \left(\hat{\psi}_i^{\dagger} \hat{V}_{ij} \hat{\psi}_j + \text{h.c.} \right) + \text{const.} , \tag{9.48}$$

where, for the case of real Hamiltonian parameters, $\hat{h}_i = (\varepsilon_i - \mu)\tau_z + \Delta_i\tau_x$ and $\hat{V}_{ij} = V_{ij}\tau_z$, $\tau_{x,z}$ being Pauli matrices in Nambu space. Notice that in this transformation there appears a residual constant which does not play any role in the system dynamics but can be relevant if one is interested in the total energy.

In addition, Keldysh–Nambu GFs can be defined as

$$\hat{G}_{ij}(t, t') = -i\langle\mathbf{T}_{\mathcal{C}}\hat{\psi}_i(t)\hat{\psi}_j^\dagger(t')\rangle \equiv -i\begin{pmatrix} \langle\mathbf{T}_{\mathcal{C}}c_{i\uparrow}(t)c_{j\uparrow}^\dagger(t')\rangle & \langle\mathbf{T}_{\mathcal{C}}c_{i\uparrow}(t)c_{j\downarrow}(t')\rangle \\ \langle\mathbf{T}_{\mathcal{C}}c_{i\downarrow}^\dagger(t)c_{j\uparrow}^\dagger(t')\rangle & \langle\mathbf{T}_{\mathcal{C}}c_{i\downarrow}^\dagger(t)c_{j\downarrow}(t')\rangle \end{pmatrix},$$

$$(9.49)$$

where one can identify electron and hole propagators (diagonal elements in Nambu space) and anomalous propagators (non-diagonal elements), associated with a finite pairing amplitude. It should be noticed that we are focusing in a spin symmetric situation where only singlet pairing is allowed. In more general situations (e.g., in the presence of magnetic fields or spin-orbit interactions) it could be necessary to go into a bi-spinor description including both spins for electron and hole components.

9.3.2 N/S Channel: Andreev Reflection

In order to illustrate the formalism let us first consider the case of a single channel contact between a normal (N) and a superconducting (S) region, which can be described using the same type of 1D models as in the previous sections but including superconducting correlations. As sketched in Fig. 9.4a, such an interface allows for novel transport processes, in addition to the normal electron transmission and reflection. An incident electron from the N region can be reflected as a hole in a process known as *Andreev* reflection. Such process should be accompanied by the creation of a Cooper pair in the S lead in order to conserve the total charge.

To get microscopic insight on these processes, the current operator of Eq. (9.26) can be written in terms of Nambu spinors as

$$\hat{I}_{LR} = \frac{ie}{\hbar}\left[\hat{\psi}_L^\dagger\hat{V}'\tau_z\hat{\psi}_R - \text{h.c.}\right],$$

$$(9.50)$$

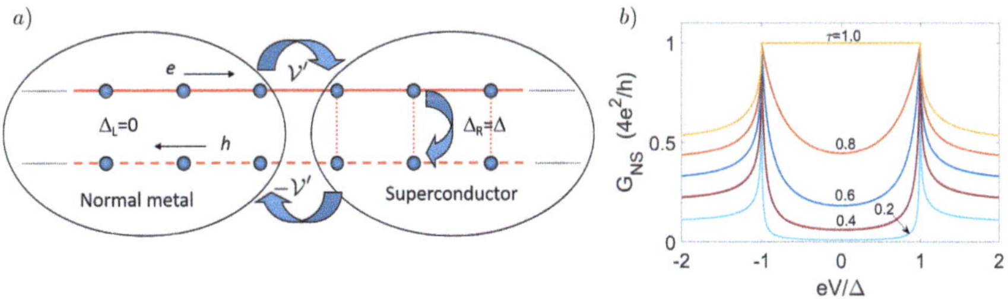

Fig. 9.4 Panel **a**: single channel model of a N/S contact exhibiting the Nambu structure. Panel **b**: differential conductance for this model at zero temperature for varying values of the normal transmission τ

so that in terms of Keldsyh–Nambu GFs we have

$$\langle \hat{I}_{LR}(t)\rangle = \frac{ie}{\hbar}\mathrm{Tr}\left[\hat{V}'\tau_z\left(\hat{G}^{+-}_{RL}(t,t) - \hat{G}^{+-}_{LR}(t,t)\right)\right]$$
$$= \frac{e}{h}\int d\omega\,\mathrm{Tr}\left[\hat{V}'\tau_z\left(\hat{G}^{+-}_{RL}(\omega) - \hat{G}^{+-}_{LR}(\omega)\right)\right] . \tag{9.51}$$

Thus, in order to obtain $\langle \hat{I}_{LR}(t)\rangle$ we would need to determine the relevant GFs component by solving the Dyson equations

$$\check{\hat{G}}^{+-} = \left(\check{\hat{I}} + \check{\hat{G}}^R\check{\hat{\Sigma}}^R\right)\check{\hat{g}}^{+-}\left(\check{\hat{I}} + \check{\hat{\Sigma}}^A\check{\hat{G}}^A\right)$$
$$\check{\hat{G}}^{R,A} = \check{\hat{g}}^{R,A}\left(\check{\hat{I}} + \check{\hat{\Sigma}}^{R,A}\check{\hat{G}}^{R,A}\right) , \tag{9.52}$$

which now include the Nambu (denoted by $\hat{\ }$) as well as the left-right (denoted by $\check{\ }$) degrees of freedom. In these equations $\check{\hat{\Sigma}}^{R,A} = V'\tau_z\sigma_x$, where now σ_x operates in the left-right space. As usual, the seeds for determining the coupled system GFs are the uncoupled GFs $\check{\hat{g}}^{+-,A,R}$, which should be diagonal in left-right space. If we choose the left to be the N lead then $\hat{g}^{+-,A,R}_L \equiv \hat{g}^{+-,A,R}_N$ should be diagonal also in Nambu space.

In order to obtain $\hat{g}^{A,R}_N$ we can extend the result in Exercise **II.6** to the Nambu space to show

$$\hat{g}^{A,R}_N(\omega) = \mathrm{diag}\left(g^{A,R}_e,\ g^{A,R}_h(\omega)\right) ,$$
$$g^{A,R}_{e,h} = \frac{1}{V}\left[\frac{\omega\mp(\varepsilon-\mu)}{2V} \pm i\sqrt{1 - \left(\frac{\omega\mp(\varepsilon-\mu)}{2V}\right)^2}\right] . \tag{9.53}$$

This expressions can be further simplified in the low energy limit (i.e., for $|\omega|$, $|\varepsilon - \mu| \ll V$), where e-h symmetry can be safely assumed and $g^{A,R}_{e,h} \simeq \pm i/V$, so that $\hat{g}^{A,R}_N \simeq \pm i\tau_0/V$. On the other hand, for $\hat{g}^{+-}_N$ we have to take into account that the chemical potential μ shifts differently for electrons and holes (i.e., it is $-\mu$ for holes) so that in the low energy approximation we have

$$\hat{g}^{+-}_N \simeq \frac{2i}{V}\mathrm{diag}\left(n_F(\omega - \mu),\ n_F(\omega + \mu)\right) . \tag{9.54}$$

In contrast to the normal case, in the superconducting lead further structure at low energy appears which can not be neglected. We can first obtain $\hat{g}^{A,R}_R \equiv \hat{g}^{A,R}_S$ from

$$\hat{g}^{A,R}_S(\omega) = \left[\omega\hat{I} - \hat{h} - \hat{V}g^{A,R}(\omega)\hat{V}\right]^{-1} , \tag{9.55}$$

with $\hat{h} = (\varepsilon - \mu)\tau_z + \Delta\tau_x$ and $\hat{V} = V\tau_z$. Further assuming e-h symmetry, i.e., $\varepsilon = \mu$, one obtains (Vecino et al. 2001) (Exercise **II.10**)

$$\hat{g}^{A,R}(\omega) = \begin{pmatrix} g^{A,R}(\omega) & f^{A,R}(\omega) \\ f^{A,R}(\omega) & g^{A,R}(\omega) \end{pmatrix} , \tag{9.56}$$

where $g^{A,R}(\omega), f^{A,R}(\omega) = \lim_{\eta\to 0} g(\omega \mp i\eta), f(\omega \mp i\eta)$, with

$$g(z) = \frac{z}{2V^2}\left[1 - \sqrt{\frac{4V^2 + \Delta^2 - z^2}{\Delta^2 - z^2}}\right], \quad f(z) = -\frac{\Delta}{z}g(z) .$$

These expressions adopt a particularly simple form in the low energy limit, i.e., for $|\omega|, \Delta \ll V$

$$g(z) \simeq -\frac{z}{V\sqrt{\Delta^2 - z^2}}, \quad f(z) \simeq \frac{\Delta}{V\sqrt{\Delta^2 - z^2}} ,$$

from which we can extract the well-known expression for the DOS within the BCS state

$$\rho_{\text{BCS}}(\omega) = \text{Re}\left[\frac{|\omega|}{\pi V\sqrt{\Delta^2 - \omega^2}}\right] , \tag{9.57}$$

exhibiting a gap for $|\omega| \leq \Delta$. In the same way $\text{Im}\, f^A(\omega) \simeq -\pi\Delta\rho_{\text{BCS}}(\omega)/\omega$ describes the spectral density for the pairing amplitude within the BCS theory.

Exercise II.10: Boundary GF for a BCS lead

Show that the retarded and advanced Green functions for a semi-infinite 1D BCS superconductor can be written as

$$\hat{g}_S^{R,A}(\omega) = \lim_{\eta\to 0} \hat{g}_S(\omega \pm i\eta)$$

where $\hat{g}_S = g\hat{\sigma}_0 + f\hat{\sigma}_x$, $\hat{\sigma}_\alpha$ are Pauli matrices in Nambu space and

$$g(\omega) = \frac{\omega}{2V^2}\left(1 - \sqrt{\frac{4V^2 + \Delta^2 - \omega^2}{\Delta^2 - \omega^2}}\right) , \quad f(\omega) = -\frac{\Delta}{\omega}g(\omega).$$

Verify that in the low energy limit, i.e., $|\omega|, \Delta \ll v$ these expressions simplify to

$$g(\omega) = \frac{-\omega}{|V|\sqrt{\Delta^2 - \omega^2}} , \quad f(\omega) = \frac{\Delta}{|V|\sqrt{\Delta^2 - \omega^2}}.$$

We finally need to determine $\hat{g}_S^{+-}(\omega)$. For this purpose it is convenient to set $\mu_s = 0$, in which case $\hat{g}_S^{+-}(\omega)$ is simply given by $n_F(\omega)\left(\hat{g}_S^A(\omega) - \hat{g}_S^R(\omega)\right)$, i.e.:

$$\hat{g}_S^{+-}(\omega) = 2i n_F(\omega)\left(\mathrm{Im}\,g^A(\omega)\tau_0 + \mathrm{Im}\,f^A(\omega)\tau_x\right) \tag{9.58}$$

In order to evaluate Eq. (9.51) for the NS case it is convenient to first simplify it using the relations

$$\hat{G}_{RL}^{+-} = \left(\hat{G}_{RR}\hat{V}'\hat{g}_L\right)^{+-} = \hat{G}_{RR}^{+-}\hat{V}'\hat{g}_L^A + \hat{G}_{RR}^R\hat{V}'\hat{g}_L^{+-}$$
$$\hat{G}_{LR}^{+-} = \left(\hat{g}_L\hat{V}'\hat{G}_{RR}\right)^{+-} = \hat{g}_L^{+-}\hat{V}'\hat{G}_{RR}^A + \hat{g}_L^R\hat{V}'\hat{G}_{RR}^{+-}\,, \tag{9.59}$$

which allows us to write (Exercise **II.11**ii)

$$\langle\hat{I}_{LR}\rangle = \frac{e\hat{V}'}{h}\int d\omega \mathrm{Tr}\left[\hat{G}_{RR}^{-+}\hat{V}'\hat{g}_L^{+-} - \hat{G}_{RR}^{+-}\hat{V}'\hat{g}_L^{-+}\right]\,, \tag{9.60}$$

and in the $T = 0$, $eV < \Delta$ limit (Exercise **II.11**iii)

$$\langle\hat{I}_{LR}\rangle = \frac{8e}{h}\frac{V'^4}{V^2}\int_{-eV}^{eV} d\omega \left|\left(\hat{G}_{RR}(\omega)\right)_{12}\right|^2\,. \tag{9.61}$$

Additionally, in the $eV \to 0$ limit one obtains the following expression for the NS linear conductance (Exercise **II.11**iv)

$$G_{NS} = \frac{4e^2}{h}R_A \tag{9.62}$$

where $R_A = \tau^2/(2-\tau)^2$ is the Andreev reflection coefficient, expressed in terms of the normal transmission coefficient τ. These results can alternatively be obtained using a scattering approach, as first shown by Blonder et al. (1982) (see also Beenakker 1992).

It should be noticed that the maximum conductance for a NS channel, reached for $\tau = 1$, is $4e^2/h$ instead of $2e^2/h$, which is due to the transfer of 2 quasiparticles for perfect Andreev reflection. In the same way, one can show that the Fano factor $S_{NS}(0)/2\langle I\rangle_{NS}$ between the zero-frequency noise and the mean current in the zero-bias, zero-temperature limit, is doubled with respect to the one reached in a normal channel, i.e., $S_{NS}(0)/2\langle I\rangle_{NS} \to 2e(1 - R_A)$.

Using the Keldysh–Nambu formalism one can also show that, for $T = 0$ (Cuevas et al. 1996)

$$G_{NS}(V) = \frac{4e^2}{h}\begin{cases} \dfrac{\tau^2}{(2-\tau)^2 - 4(1-\tau)\left(\frac{eV}{\Delta}\right)^2} & eV \le \Delta \\[3ex] \dfrac{\tau}{\tau + (2-\tau)\sqrt{1-\left(\frac{\Delta}{eV}\right)^2}} & eV > \Delta \end{cases}\,, \tag{9.63}$$

which exhibits the behavior illustrated in Fig. 9.4b.

> **Exercise II.11: Andreev reflection**
>
> Using the single channel model for a NS junction
>
> (i) Show that the expression for the mean current in the Nambu formalism can be written
>
> $$\langle \hat{I}_{LR} \rangle = \frac{e}{h} \int d\omega \mathrm{Tr} \left[\sigma_z \left(\hat{V}' \hat{G}_{RL}^{+-} - \hat{V}' \hat{G}_{RL}^{+-} \right) \right] ,$$
>
> where $\hat{V}' = V' \hat{\sigma}_z$.
>
> (ii) From this expression and taking into account that $L \equiv N$ and $R \equiv S$ show that
>
> $$\langle \hat{I}_{LR} \rangle = \frac{eV'}{h} \int d\omega \mathrm{Tr} \left[\hat{G}_{RR}^{-+} \hat{V}' \hat{g}_L^{+-} - \hat{G}_{RR}^{+-} \hat{V}' \hat{g}_L^{-+} \right].$$
>
> (iii) Show that in the zero temperature/zero voltage limit, the linear conductance for the NS junction is given by
>
> $$G_{NS} = \frac{4e^2}{h} R_A$$
>
> where
>
> $$R_A = \frac{4\hat{V}'^4}{V^2} | \left(\hat{G}_{RR}^A \right)_{12} |^2$$
>
> (iv) Finally use the low energy approximation for the leads GFs to show that
>
> $$R_A = \frac{\tau^2}{(2 - \tau)^2}$$
>
> where τ is the normal transmission coefficient found in Exercise **II.7**.

9.3.3 *S/S Channel: Josephson Effect*

When both leads are superconducting electron transport becomes more involved due to the emergence of the Josephson effect (JE), which is a paradigmatic manifestation of quantum coherence at the macroscopic scale. Let us first focus on the *dc* JE by noticing that the mean field solution of the BCS Hamiltonian contains a degeneracy associated to the phase ϕ of the order parameter Δ (such that $\Delta = |\Delta| e^{i\phi}$). It can be shown that the ground state energy of (9.44) does not depend on ϕ while the corresponding eigenstates $|\Phi_\phi\rangle$ with different ϕ are linearly independent solutions.

In fact ϕ corresponds to the conjugate variable to the number of Cooper pairs, such that a state with a fixed number of pairs $|\Phi_N\rangle$ can be written as

$$|\Phi_N\rangle = \frac{1}{2\pi} \int_0^{2\pi} d\phi e^{iN\phi} |\Phi_\phi\rangle \; . \tag{9.64}$$

While ϕ is irrelevant for a bulk system or in the previously discussed NS case, this is not the case for a S/S junction, as demonstrated by Josephson (1962). He showed that in a tunnel junction between two S leads characterized by a superconducting phase difference $\phi = \phi_L - \phi_R$ there appears a non-dissipative current (i.e., flowing in the absence of an applied bias voltage) given by $I_c \sin\phi$, where I_c determines the maximum supercurrent.

We can use our single-channel model to analyze how this effect evolves as a function the channel transmission. For that purpose we should evaluate Eq. (9.51) assuming that now L/R correspond to two S leads with $\Delta_{L,R} = \Delta e^{i\phi_{L,R}}$. First notice that the dc JE is an equilibrium phenomenon so that

$$\hat{G}_{ij}^{+-} = n_F(\omega) \left(\hat{G}_{ij}^A(\omega) - \hat{G}_{ij}^R(\omega) \right) \; . \tag{9.65}$$

On the other hand, it is convenient to remove the phase from the S leads by a unitary transformation $c_{i\sigma} \to \tilde{c}_{i\sigma} e^{-i\phi_i/2}$ so that it enters just into the $\hat{V}'$ matrix, i.e., $\hat{V}' \to V' \tau_z e^{i\phi\tau_z/2}$. It is then straightforward to show from Eq. (9.51) that (Exercise **II.12**a)

$$\langle \hat{I}_{LR} \rangle = \frac{2e}{h} \int d\omega n_F(\omega) \mathrm{ReTr} \left[\tau_z \left\{ \hat{V}' \hat{G}_{RL}^A(\omega) - \hat{V}'^\dagger \hat{G}_{LR}^A(\omega) \right\} \right] \; . \tag{9.66}$$

Further manipulations allows one to show that Martín-Rodero et al. (1994) (Exercise **II.12**b)

$$\langle \hat{I}_{LR} \rangle = \frac{8eV'^2}{h} \sin\phi \int d\omega n_F(\omega) \mathrm{Im} \left(\frac{f^A(\omega) f^A(\omega)}{D^A(\omega)} \right) \; , \tag{9.67}$$

where $D^A(\omega) = \det \left[\hat{I} - \hat{V}'^\dagger \hat{g}_L^A(\omega) \hat{V}' \hat{g}_R^A(\omega) \right]$. (For this purpose one can use the Jacobi's formula defining $\hat{A}(\phi, \omega) = \left[\hat{I} - \hat{V}'^\dagger \hat{g}_L^A(\omega) \hat{V}' \hat{g}_R^A(\omega) \right]$). Finally, in the zero temperature limit and using the low energy approximation for the $g's$, the integral in Eq. (9.67) can be solved by residues, leading to (Exercise **II.12**c)

$$\langle \hat{I}_{LR} \rangle = \frac{e\Delta}{2\hbar} \frac{\tau \sin\phi}{\sqrt{1 - \tau \sin^2 \frac{\phi}{2}}} \; . \tag{9.68}$$

A nice feature of Eq. (9.68) is that it describes both the $\tau \to 0$ limit, where $\langle \hat{I}_{LR} \rangle \sim (e\Delta\tau/2\hbar)\sin\phi$ as expected for a tunnel junction, while for $\tau \to 1$ it leads to $\langle \hat{I}_{LR} \rangle \sim (e\Delta/\hbar)\sin(\phi/2)$, as first predicted by Kulik and Omelyanchuk (1978).

Exercise II.12: Supercurrent and Andreev bound states

(i) Demonstrate that in the single channel model the mean current for a S/S junction of arbitrary transmission is given by

$$\langle I_S \rangle = \frac{8e\mathcal{V}'^2}{h}\sin\phi \int d\omega\, n_F(\omega)\,\mathrm{Im}\left[\frac{f_L^A(\omega)f_R^A(\omega)}{D^A(\omega)}\right],$$

where ϕ is the superconducting phase difference and

$$D^A(\omega) = \det\left[\hat{I} - \hat{\mathcal{V}}'^{\dagger}\hat{g}_L^A \hat{v}' \hat{g}_R^A\right]$$

where $\hat{\mathcal{V}}' = \mathcal{V}'\hat{\sigma}_z e^{i\hat{\sigma}_z\phi/2}$.

(ii) Show that the integrand in the above equation has poles at the energies

$$\omega_{1,2}(\phi) = \pm\Delta\sqrt{1 - \tau\sin^2\frac{\phi}{2}}$$

(iii) (Optional) Use the above result to show that at zero temperature

$$\langle I_S \rangle = \frac{e\Delta}{2\hbar}\frac{\tau\sin\phi}{\sqrt{1 - \tau\sin^2\frac{\phi}{2}}}$$

9.3.4 Transport Through Majorana Nanowires

In this section we illustrate the use of the Keldysh–Nambu formalism for describing transport through Majorana nanowire junctions. We consider the simplest model for a topological superconductor (TS) which is provided by the spinless Kitaev model in 1D (Kitaev 2001). It can be defined in terms of Nambu spinors $\hat{\Psi}_k = \left(c_k,\ c_{-k}^{\dagger}\right)^T$ as

$$H_K = \frac{1}{2}\sum_k \hat{\Psi}_k^{\dagger}\hat{h}_k\hat{\Psi}_k \sum_k = \frac{1}{2}\hat{\Psi}_k^{\dagger}\left(-\mathcal{V}\cos(k)\tau_z + \Delta\sin(k)\tau_y\right)\hat{\Psi}_k, \tag{9.69}$$

where $\mathcal{V}$ and Δ denote the hopping and pairing amplitudes respectively.

As in the normal case, the real space GFs corresponding to this model can be expressed as a contour integral in the $z = e^{ik}$ plane

$$\hat{\mathcal{G}}_{ij}(\omega) = \oint_{|z|=1} \frac{dz}{2\pi i} \left(\omega \pm i0^+ - \hat{h}_k \right)^{-1} z^{i-j-1} . \tag{9.70}$$

In order to obtain the boundary GFs for this model, $\hat{g}_{TS}$, we only need the $i, j \in \{0, \pm 1\}$ components. From Eq. (9.70), one obtains (Zazunov et al. 2016)

$$\hat{\mathcal{G}}_{00}(\omega) = \frac{-2\omega\tau_0}{\sqrt{(\omega^2 - \Delta^2)(\omega^2 - 4V^2)}}, \tag{9.71}$$

$$\hat{\mathcal{G}}_{\pm 1,0}(\omega) = \hat{\mathcal{G}}_{0,\mp 1}(\omega) = \frac{2V(z_1^2 + 1)\tau_z \pm i\Delta(z_1^2 - 1)\tau_y}{\sqrt{(\omega^2 - \Delta^2)(\omega^2 - 4V^2)}},$$

where

$$z_1^2 = \frac{2\omega^2 - (4V^2 + \Delta^2)}{4V^2 - \Delta^2} - \mathrm{sgn}\left(2\omega^2 - (4V^2 + \Delta^2)\right) \sqrt{\left(\frac{2\omega^2 - (4V^2 + \Delta^2)}{4V^2 - \Delta^2}\right)^2 - 1}.$$

Using these expressions and proceeding as for the normal case but extended to the Nambu space, in the low energy limit ($|\omega|, |\Delta| \ll V$) one obtains (Exercise **II.13**a)

$$\hat{g}_{TS}^{R/A}(\omega) = \frac{\sqrt{\Delta^2 - (\omega \pm i0^+)^2}\,\tau_0 + \Delta\tau_x}{V(\omega \pm i0^+)}. \tag{9.72}$$

From Eq. (9.72) the boundary spectral density for a TS described by the Kitaev model can be obtained as the Nambu trace of

$$-\frac{1}{\pi}\mathrm{Im}\hat{g}_{TS}^R(\omega) = \frac{\Delta}{V}[\sigma_0 + \sigma_x]\delta(\omega) + \frac{\sqrt{\omega^2 - \Delta^2}}{\pi V|\omega|}\sigma_0\Theta(|\omega| - \Delta) , \tag{9.73}$$

which features the celebrated Majorana zero-energy peak due to the $\omega = 0$ pole of the retarded GF in Eq. (9.72).

As discussed in Zazunov et al. (2016) transport properties in several different situations can be analyzed based on Eqs. (9.72) and (9.73). Particularly simple is the case of a N/TS junction, where one obtains (Exercise **II.13**b)

$$I = \frac{e}{h} \int d\omega \left[n_F(\omega - eV) - n_F(\omega + eV)\right] J(\omega), \tag{9.74}$$

where the current density $J(\omega)$ adopts the rather simple form

$$J(\omega) = \begin{cases} 1/(1 + \omega^2/\Gamma^2), & |\omega| < \Delta, \\[2ex] \tau\dfrac{\tau + (2-\tau)\sqrt{1 - (\Delta/\omega)^2}}{\left[2 - \tau + \tau\sqrt{1 - (\Delta/\omega)^2}\right]^2}, & |\omega| \geq \Delta, \end{cases} \tag{9.75}$$

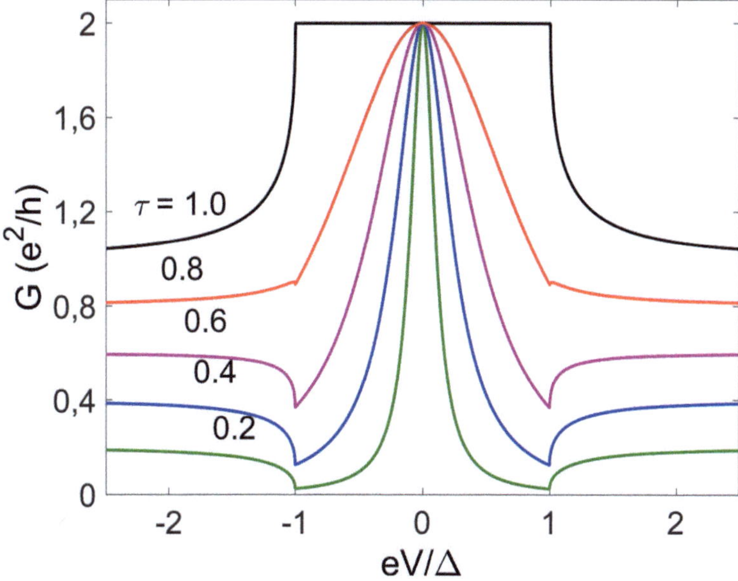

Fig. 9.5 Differential conductance for a N/TS junction, as obtained from Eq. (9.74) at zero temperature

where τ denotes the usual normal transmission coefficient and the rate Γ is given by

$$\Gamma = \frac{\tau\Delta}{2\sqrt{1-\tau}}. \tag{9.76}$$

The results for the differential conductance obtained from this model are illustrated in Fig. 9.5.

Another simple application of the formalism is provided by Josephson transport in a TS/TS junction. In this case one can show (Exercise **II.13**c) that the Andreev bound states are determined by

$$\omega_{1,2}(\phi) = \pm\sqrt{\tau}\Delta\cos(\phi/2) \,. \tag{9.77}$$

The resulting 4π periodicity of the Josephson effect, as a possible signature of Majorana bound states, has been the object of intense debate in the literature, but is out of the scope of this book.

Exercise II.13:   Transport in Majorana junctions

(i) From the real space GFs for the bulk Kitaev model given in Eq. (9.71) obtain the boundary GF for this model in the low energy approximation (c.f. (9.72)).

(ii) Derive the expression of the spectral current $J(\omega)$ in a N/TS junction given in Eq. (9.75).

(iii) Demonstrate that the Andreev bound states in a TS/TS Josephson junction are given by Eq. (9.77).

Part III
Path Integral Formulation of the Quantum Many-Body Problem

Chapter 10
Introduction to Path Integral Methods

Path integrals were introduced by Feynman as an alternative formulation to the conventional hamiltonian approach used in quantum mechanics. The quantum many-body problem can also be formulated in an alternative way without using non-commuting operators (as we have been doing up to here). The partition function, correlation functions and Green functions can be obtained in terms of path integrals. This is in close analogy with classical statistical mechanics. The quantum problem in d-dimensions can be formulated as a classical system with dimensions, $d + 1$. Hence, the concepts of critical phenomena scaling and renormalization group of classical systems can be applied to quantum systems. We would like to use second quantization language to express path integrals. Coherent states are the natural states to express the path integral with numbers (instead of c, $c^\dagger$ operators). These numbers are eigenstates of the c operator. The c and $c^\dagger$ operators play the same role as the position x and p operators in standard quantum mechanics.

10.1 Path Integral of a Single Boson

Let's consider the simplest hamiltonian of a single boson:

$$H = \omega b^\dagger b \tag{10.1}$$

with: $[b, b^\dagger] = 1$; $[b, b] = [b^\dagger, b^\dagger] = 0$.

Which is the path integral representation of a single boson? To do this we need boson coherent states:

$$b|\phi\rangle = \phi|\phi\rangle$$
$$\langle\bar{\phi}|b^\dagger = \langle\bar{\phi}|\bar{\phi} \tag{10.2}$$

where ϕ and $\bar{\phi}$ are complex numbers and $\{|\phi\rangle, |\bar{\phi}\rangle\}$.

© The Author(s), under exclusive license to Springer Nature Switzerland AG 2024
J. Merino and A. L. Yeyati, *Many-Body Techniques in Condensed Matter Physics*,
UNITEXT for Physics, https://doi.org/10.1007/978-3-031-55143-7_10

The coherent state of a single boson operator can be expressed as:

$$|\phi\rangle = e^{\phi b^\dagger}|0\rangle = (1 + \phi b^\dagger + \frac{(\phi b^\dagger)^2}{2!} + \cdots)|0\rangle = |0\rangle + \phi|1\rangle + \frac{\phi^2}{2!}|2\rangle + \cdots$$

$$(10.3)$$

which is a linear combination of all Fock states (which form a complete basis). Hence coherent states form an overcomplete basis.

We can check that $\{|\phi\rangle\}$ is indeed an eigenstates of b:

$$b|\phi\rangle = b(e^{\phi b^\dagger}|0\rangle) = b|0\rangle + \phi b b^\dagger|0\rangle + \frac{\phi^2}{2!}bb^\dagger b^\dagger|0\rangle + \ldots$$

$$= \phi|0\rangle + \frac{\phi^2}{2!}(1 + b^\dagger b)b^\dagger|0\rangle + \ldots$$

$$= \phi(1 + \phi b^\dagger + \frac{(\phi b^\dagger)^2}{2!} + \cdots)|0\rangle = \phi|\phi\rangle. \qquad (10.4)$$

Thus:

$$b|\phi\rangle = \phi|\phi\rangle, \; \langle\bar\phi|b^\dagger = \langle\bar\phi|\bar\phi \qquad (10.5)$$

Also:

$$\langle\bar\phi| = \langle 0|e^{b\bar\phi} \qquad (10.6)$$

A coherent state describes a condensate with an indefinite number of particles since:

$$|\phi\rangle = \sum_n \frac{\phi^n}{n!}(b^\dagger)^n|0\rangle = \sum_n \frac{\phi^n}{\sqrt{n!}}|n\rangle \qquad (10.7)$$

where:

$$|n\rangle = \frac{(b^\dagger)^n}{\sqrt{n!}}|0\rangle. \qquad (10.8)$$

So the amplitude for a coherent state to have n particles is:

$$\langle n|\phi\rangle = \frac{\phi^n}{\sqrt{n!}}. \qquad (10.9)$$

Similarly:

$$\langle\bar\phi| = \sum_m \frac{\bar\phi^m}{\sqrt{m!}}\langle m| \qquad (10.10)$$

and:

$$\langle\bar\phi|m\rangle = \frac{\bar\phi^m}{\sqrt{m!}}. \qquad (10.11)$$

The overlap between two coherent states is:

$$\langle \bar{\phi} | \psi \rangle = \sum_{m,n} \frac{\bar{\phi}^m}{\sqrt{m!}} \langle m | n \rangle \frac{\psi^n}{\sqrt{n!}} = \sum_n \frac{(\bar{\phi}\psi)^n}{n!} = e^{\bar{\phi}\psi} \neq 0 \tag{10.12}$$

so two coherent states are not orthogonal! Even though coherent states are not orthogonal they can still be used as a convenient basis to express the $b, b^\dagger$ since they become diagonal in such basis. We can obtain matrix elements easily:

$$\langle \bar{\phi} | : O[b^\dagger, b] : | \psi \rangle = O[\bar{\phi}, \psi] \langle \bar{\phi} | \psi \rangle = O[\bar{\phi}, \psi] e^{\bar{\phi}\psi}. \tag{10.13}$$

for any normal ordered operator. The unit operator can be decomposed in terms of coherent states as:

$$\int \frac{d\bar{\phi} d\phi}{-2\pi i} e^{-\bar{\phi}\phi} | \phi \rangle \langle \bar{\phi} | = 1 \tag{10.14}$$

*Demonstrate this. Note that the integration can be expressed as:

$$\int \frac{d(Re\phi)d(Im\phi)}{\pi} = \int \frac{d\bar{\phi} d\phi}{-2\pi i} \tag{10.15}$$

An example of a boson coherent state is a laser beam. A laser beam is a coherent state of photons which leads to the electromagnetic field: $\langle E \rangle \propto e^{i(kz-\omega t)}$. It can be understood as a condensate with an indefinite number of photons.

The average photon number in a coherent state $|\phi\rangle$ is:

$$\bar{n} = \langle n \rangle = \frac{\langle \bar{\phi} | b^\dagger b | \phi \rangle}{\langle \bar{\phi} | \phi \rangle} = |\phi|^2 \frac{\langle \bar{\phi} | \phi \rangle}{\langle \bar{\phi} | \phi \rangle} = |\phi|^2 \tag{10.16}$$

The variance in the distribution of the number of particles in a coherent state is:

$$\frac{\sigma}{n} = \sqrt{\frac{\sigma^2}{\bar{n}^2}} = \frac{1}{\bar{n}} \sqrt{\frac{\langle \bar{\phi} | n^2 | \phi \rangle}{\langle \bar{\phi} | \phi \rangle} - \left(\frac{\langle \bar{\phi} | n | \phi \rangle}{\langle \bar{\phi} | \phi \rangle} \right)^2} = \frac{1}{\bar{n}} \sqrt{\bar{n}^2 + \bar{n} - \bar{n}^2} = \frac{1}{\sqrt{\bar{n}}} \to 0, \text{ as } \bar{n} \to \infty \tag{10.17}$$

The probability $p(n)$ of a coherent state $|\phi\rangle$ to be in a state with n particles:

$$p(n) = \frac{1}{n!} (\bar{\phi}\phi)^n e^{-\phi\bar{\phi}} \tag{10.18}$$

is a Poisson distribution which can be shown from the relation:

$$p(n) = \frac{|\langle n | \phi \rangle|^2}{\langle \bar{\phi} | \phi \rangle} \tag{10.19}$$

In the thermodynamic limit it follows from Eq. (10.17) that the Poisson distribution of the particle number in a coherent state becomes a Dirac delta function.

Partition function

We can now evaluate the partition function of a single boson by using the coherent state basis introduced above. The partition function reads:

$$Z = Tr\{e^{-\beta H}\} = \sum_n \langle n|e^{-\beta H}|n\rangle, \tag{10.20}$$

with: $\sum_n |n\rangle\langle n| = 1$, with $\{|n\rangle\}$ being any complete set of eigenstates. We can insert the identity expressed in terms of coherent states of bosons:

$$Z = \sum_n \int \frac{d\bar{\phi}d\phi}{(-2\pi i)} e^{-\bar{\phi}\phi} \langle n|\phi\rangle\langle\bar{\phi}|e^{-\beta H}|n\rangle = \int \frac{d\bar{\phi}d\phi}{(-2\pi i)} e^{-\bar{\phi}\phi} \langle\bar{\phi}|e^{-\beta H}|\phi\rangle, \tag{10.21}$$

where H is in normal ordered form so $e^{-\beta :H:} =: e^{-\beta H} :$. We can compute the $\langle\bar{\phi}|e^{-\beta H}\phi\rangle$ as a path integral by breaking the 0 to β interval in M steps:

$$\langle\bar{\phi}|e^{-\beta H}|\phi\rangle = \langle\bar{\phi}(e^{-\epsilon :H:})^M|\phi\rangle = \langle\bar{\phi}|(e^{-\beta :H:})^M|\phi\rangle = \langle\bar{\phi}| : e^{-\epsilon H} : I : e^{-\epsilon H} : I : e^{-\epsilon H} : \ldots |\phi\rangle.$$

In the breakup of the time integration, $\phi(\tau) \to (\phi_0, \phi_1, \ldots \phi_M)$, $\bar{\phi}(\tau) \to (\bar{\phi}_0, \bar{\phi}_1, \ldots \bar{\phi}_M)$ we associate: $|\phi\rangle = |\phi_0\rangle$ and $|\bar{\phi}\rangle = |\bar{\phi}_M\rangle$. Expressing the intermediate identities as coherent state path integration over coherent states we have:

$$\langle\bar{\phi}|e^{-\beta H}|\phi\rangle = \lim_{M\to\infty} \int \left(\frac{d\bar{\phi}_{M-1}\ldots d\phi_{M-1}}{-2\pi i}\right)\left(\frac{d\bar{\phi}_{M-2}\ldots d\phi_{M-2}}{-2\pi i}\right)\cdots\left(\frac{d\bar{\phi}_1\ldots d\phi_1}{-2\pi i}\right)$$
$$\times \langle\bar{\phi}_M| : e^{-\epsilon H} : |\phi_{M-1}\rangle\langle\bar{\phi}_{M-1}| : e^{-\epsilon H} : |\phi_{M-2}\rangle\langle\bar{\phi}_{M-2}| : e^{-\epsilon H} : |\phi_{M-3}\rangle \ldots |\phi_1\rangle\langle\bar{\phi}_1| : e^{-\epsilon H} : |\phi_0\rangle$$
$$\times e^{-\bar{\phi}_1\phi_1}e^{-\bar{\phi}_2\phi_2}\ldots e^{-\bar{\phi}_{M-1}\phi_{M-1}}. \tag{10.22}$$

Hence:

$$\langle\bar{\phi}|e^{-\beta H}|\phi\rangle = \lim_{M\to\infty} \int \prod_{j=1}^{M-1} \frac{d\bar{\phi}_j d\phi_j}{(-2\pi i)} e^{-\sum_{j=1}^{M-1}\bar{\phi}_j\phi_j} \prod_{k=1}^{M}\langle\bar{\phi}_k| : e^{-\epsilon H} : |\phi_{k-1}\rangle. \tag{10.23}$$

We can now evaluate the imaginary time evolution during a stepsize ϵ:

$$\langle\bar{\phi}_k| : e^{-\epsilon H} : |\phi_{k-1}\rangle = \langle\bar{\phi}_k|\phi_{k-1}\rangle - \epsilon\langle\bar{\phi}_k| : H : |\phi_{k-1}\rangle$$
$$= \langle\bar{\phi}_k|\phi_{k-1}\rangle - \epsilon\langle\bar{\phi}_k|\phi_{k-1}\rangle H(\bar{\phi}_k, \phi_{k-1})$$
$$= e^{\bar{\phi}_k\phi_{k-1}}(1 - \epsilon H(\bar{\phi}_k, \phi_{k-1})) = e^{\bar{\phi}_k\phi_{k-1}}e^{-\epsilon H(\bar{\phi}_k,\phi_{k-1})}. \tag{10.24}$$

When inserting this result in the partition function we find:

$$Z = \lim_{M \to \infty} \int \prod_{j=1}^{M} \frac{d\bar{\phi}_j d\phi_j}{(-2\pi i)} \prod_{k=1}^{M} e^{-\epsilon \bar{\phi}_k (\frac{\phi_k - \phi_{k-1}}{\epsilon})} (1 - \epsilon H[\bar{\phi}_k, \phi_{k-1}]), \qquad (10.25)$$

where we have identified: $\phi_0 = \phi_M$, $\bar{\phi}_0 = \bar{\phi}_M$.

This can be expressed in a compact form as a functional path integration:

$$Z = \int_{\phi(\beta)=\phi(0)} \mathcal{D}\bar{\phi}(\tau) \mathcal{D}\phi(\tau) e^{-\int_0^\beta d\tau (\bar{\phi}(\tau) \frac{\partial}{\partial \tau} \phi(\tau) + H[\bar{\phi}(\tau), \phi(\tau)])} \qquad (10.26)$$

where the integral is along closed by bosonic fields which satisfy a periodic boundary condition. This is the path integral expression for the partition function of a single boson field with anormal ordered hamiltonian: $H = \omega b^\dagger b = H[b^\dagger, b] =: H[b^\dagger, b] :$.

The integral represents a sum over all possible "histories" of the discretized fields: $\phi(\tau) = (\phi_1, \phi_2, \ldots, \phi_M)$, $\bar{\phi}(\tau) = (\bar{\phi}_1, \bar{\phi}_2, \ldots, \bar{\phi}_M)$. This is a functional integral. When we identify: $\phi_0 = \phi_M$ and $\bar{\phi}_0 = \bar{\phi}_M$ it implies the periodic boundary condition over the bosonic fields: $\phi(\tau) = \phi(\tau + \beta)$, $\bar{\phi}(\tau) = \bar{\phi}(\tau + \beta)$, i.e., is periodic on time.

If we now introduce the hamiltonian of a non-interacting boson such as the harmonic oscillator, we would have:

$$Z = \int_{\phi(\beta)=\phi(0)} \mathcal{D}\bar{\phi}(\tau) \mathcal{D}\phi(\tau) e^{-\int_0^\beta d\tau \bar{\phi}(\tau)(\frac{\partial}{\partial \tau} + \omega)\phi(\tau)} \qquad (10.27)$$

For a multiple boson hamiltonian like:

$$H = \sum_\lambda b_\lambda^\dagger b_\lambda, \qquad (10.28)$$

with the general form: $H[b_\lambda^\dagger, b_\lambda] =: H[b_\lambda^\dagger, b_\lambda] :$, the path integral generalizes to:

$$Z = \int_{\phi(\beta)=\phi(0)} \mathcal{D}\bar{\phi}(\tau) \mathcal{D}\phi(\tau) e^{-\int_0^\beta d\tau \sum_\lambda (\bar{\phi}_\lambda(\tau) \frac{\partial}{\partial \tau} \phi_\lambda(\tau) + H[\bar{\phi}_\lambda(\tau), \phi_\lambda(\tau)])}$$

This would apply for the free boson gas, i.e., when $\lambda = \mathbf{k}$.

10.1.1 Green Function from Path Integration

Apart from the partition function, the path integral approach can also be used to obtain time-ordered expectation values. The division of the time variable into M time-slices using coherent states can be carried out for the evaluation of arbitrary time-ordered products of fields finding that these time-ordered products map onto a path integral over the c-number products of the fields. For instance, for the one-particle Green function, we have:

$$G(x - x', \tau - \tau') = -\langle T_\tau [b(x\tau) b^\dagger(x'\tau')] \rangle = -\frac{1}{Z} \int \mathcal{D}\bar{\phi}\mathcal{D}\phi \, \phi(x, \tau)\bar{\phi}(x'\tau') e^{-S},$$

(10.29)

where the partition function reads:

$$Z = \int \mathcal{D}\bar{\phi}\mathcal{D}\phi \, e^{-S},$$

(10.30)

where:

$$S = \int_0^\beta d\tau \{\bar{\phi}\partial_\tau \phi + H[\bar{\phi}, \phi]\}.$$

(10.31)

Time-ordered correlation functions such as the Green function may be obtained from the generating functional:

$$Z[\eta(x, \tau), \bar{\eta}(x, \tau)] = \int \mathcal{D}\bar{\phi}\mathcal{D}\phi \, e^{-S + \int [\bar{\eta}(x,\tau)\phi(x,\tau) + \bar{\phi}(x,\tau)\eta(x,\tau)]d\tau} = \int \mathcal{D}\bar{\phi}\mathcal{D}\phi \, e^{-S[\bar{\eta},\eta]},$$

(10.32)

from which we can obtain time-ordered products of bosonic fields:

$$G(x - x', \tau - \tau') = -\frac{\delta^2 \ln Z[\eta, \bar{\eta}]}{\delta\eta(x', \tau')\delta\bar{\eta}(x, \tau)}\Big|_{\eta,\bar{\eta}=0} = \frac{1}{Z}\frac{\delta^2 Z[\eta, \bar{\eta}]}{\delta\eta(x', \tau')\delta\bar{\eta}(x, \tau)}\Big|_{\eta,\bar{\eta}=0}$$
$$= -\frac{1}{Z}\int \mathcal{D}\bar{\phi}\mathcal{D}\phi \, \phi(x, \tau)\bar{\phi}(x'\tau') e^{-S}.$$

10.1.2   Single Boson

As an example we obtain the Green function of a single boson with hamiltonian:

$$H = \omega_0 b^\dagger b$$

(10.33)

The partition function reads:

$$Z = \int \mathcal{D}\bar{\phi}\mathcal{D}\phi \, e^{-\int_0^\beta d\tau \bar{\phi}(\tau)(\partial_\tau + \omega_0)\phi(\tau)} = \frac{1}{|\partial_\tau + \omega_0|}$$

(10.34)

Adding the sources, the generating functional is:

$$Z[\eta, \bar{\eta}] = \int \mathcal{D}\bar{\phi}\mathcal{D}\phi \, e^{-\int_0^\beta (\bar{\phi}(\partial_\tau + \omega_0)\phi - \bar{\eta}\phi - \bar{\phi}\eta)} = \frac{e^{\bar{\phi}(\partial_\tau + \omega_0)^{-1}\eta}}{|\partial_\tau + \omega_0|}. \tag{10.35}$$

The Green function can be obtained from $Z[\eta, \bar{\eta}]$ as:

$$G \equiv -\frac{1}{Z}\frac{\delta^2 Z[\eta, \bar{\eta}]}{\delta\eta\delta\bar{\eta}}\Big|_{\eta = \bar{\eta} = 0} = -\frac{1}{Z}\frac{\delta}{\delta\eta}\Big\{\frac{(\partial_\tau + \omega_0)^{-1}\eta e^{\bar{\eta}(\partial_\tau + \omega_0)^{-1}\eta}}{|\partial_\tau + \omega_0|}\Big\}_{\eta, \bar{\eta} = 0} = -(\partial_\tau + \omega_0)^{-1} \tag{10.36}$$

These expressions can be extended to any free (quadratic) bosonic hamiltonian of the type:

$$H = \sum_{\alpha\beta} b_\alpha^\dagger h_{\alpha\beta} b_\beta = b^\dagger h b \tag{10.37}$$

and so:

$$Z = \frac{1}{|\partial_\tau + h|}$$

$$Z[\bar{\eta}, \eta] = \frac{e^{\bar{\eta}(\partial_\tau + h)^{-1}\eta}}{|\partial_\tau + h|}. \tag{10.38}$$

and the Green function:

$$G = -(\partial_\tau + h)^{-1}. \tag{10.39}$$

Note on Gaussian integration The above expressions are obtained by performing Gaussian integration. Gaussian integration in the complex plane gives:

$$\int \frac{d\bar{\phi}d\phi}{-2\pi i} e^{-\bar{\phi}W\phi} = \frac{1}{W} \tag{10.40}$$

In the case of multidimensional Gaussian integration:

$$\int \prod_\alpha \frac{d\bar{\phi}_\alpha d\phi_\alpha}{-2\pi i} e^{-\bar{\phi}_\alpha M_{\alpha\beta}\phi_\beta} = \frac{1}{|M|}. \tag{10.41}$$

When source terms are present we have:

$$\int \prod_\alpha \frac{d\bar{\phi}_\alpha d\phi_\alpha}{-2\pi i} e^{-\bar{\phi}_\alpha M_{\alpha\beta}\phi_\beta - \bar{\eta}_\alpha \phi_\alpha - \bar{\phi}_\alpha \eta_\alpha} = \frac{e^{\bar{\eta}M^{-1}\eta}}{|M|}. \tag{10.42}$$

This result can be obtained simply from the result for the partition function making the change of variables:

$$\frac{1}{|M|} = \int \frac{d\bar{\phi}d\phi}{(-2\pi i)} e^{-(\bar{\phi}-\bar{\eta}M^{-1})M(\phi-M^{-1}\eta)} = e^{-\bar{\eta}M^{-1}\eta} \int \frac{d\bar{\phi}d\phi}{-2\pi i} e^{-\bar{\phi}M\phi} e^{\bar{\phi}\eta} e^{\bar{\eta}\phi}$$

$$(10.43)$$

Fourier transform

Since we are considering systems homogeneous in time it is convenient to perform Fourier transformation and obtain the Green functions in frequency space. The bosonic fields are periodic functions in the $(0, \beta)$ interval: $\phi(\tau) = \phi(\tau + \beta)$. Expanding in bosonic Matsubara frequencies:

$$\phi(\tau) = \frac{1}{\sqrt{\beta}} \sum_n e^{-i\omega_n \tau} \phi(\omega_n)$$

$$\bar{\phi}(\tau) = \frac{1}{\sqrt{\beta}} \sum_n e^{i\omega_n \tau} \bar{\phi}(\omega_n)$$

$$(10.44)$$

with $\omega_n = \frac{2\pi n}{\beta}$. In terms of these fields the action can be decomposed as:

$$S = \int_0^\beta d\tau (\bar{\phi}(\tau) \frac{\partial}{\partial \tau} \phi(\tau) + H[\bar{\phi}(\tau), \phi(\tau)]) = \frac{1}{\beta} \sum_{n,m} \int_0^\beta d\tau e^{i(\omega_m - \omega_n)\tau} \bar{\phi}(\omega_m)(-i\omega_n + \omega_0)\phi(\omega_n).$$

$$(10.45)$$

Using:

$$\frac{1}{\beta} \int_0^\beta d\tau e^{i(\omega_m - \omega_n)} = \delta_{mn}$$

$$(10.46)$$

the action becomes diagonal in frequency space:

$$S = \sum_n \bar{\phi}(\omega_n)(-i\omega_n + \omega_0)\phi(\omega_n),$$

$$(10.47)$$

and the partition function simplifies to:

$$Z = \int \mathcal{D}\bar{\phi}\mathcal{D}\phi\, e^{-\sum_n \bar{\phi}(\omega_n)(-i\omega_n + \omega_0)\phi(\omega_n)}.$$

$$(10.48)$$

Since Z can be factorized into a product of Gaussian integrals (one integral per Fourier component $\phi(\omega_n)$) we find:

$$Z = \frac{1}{\prod_n (-i\omega_n + \omega_0)}.$$

$$(10.49)$$

We can also obtain the propagators from the generating functional in frequency space:

$$Z[\eta, \bar{\eta}] = \int \mathcal{D}\bar{\phi}\mathcal{D}\phi\, e^{-\sum_n \bar{\phi}(\omega_n)(-i\omega_n + \omega_0)\phi(\omega_n) - \bar{\eta}(\omega_n) - \bar{\phi}(\omega_n)\eta(\omega_n)}.$$

$$(10.50)$$

Performing the functional derivatives, the Green function reads:

$$G = -\langle \phi(\omega_1)\bar{\phi}(\omega_2)\rangle \equiv -\frac{1}{Z}\frac{\delta^2 Z[\eta,\bar{\eta}]}{\delta\eta(\omega_2)\delta\bar{\eta}(\omega_1)}\Big|_{\eta=\bar{\eta}=0} = -\frac{1}{Z}\int \mathcal{D}\bar{\phi}(\omega_n)\mathcal{D}\phi(\omega_n)e^{-S}\phi(\omega_1)\bar{\phi}(\omega_2)$$

$$(10.51)$$

Since the generating functional can be expressed as:

$$Z[\eta,\bar{\eta}] = \frac{e^{\bar{\eta}M^{-1}\eta}}{|M|}, \tag{10.52}$$

with: $M = (-i\omega_n + \omega_0)I$, and the Green function:

$$G = -\langle \phi(\omega_1)\bar{\phi}(\omega_2)\rangle = \frac{-1}{-i\omega_2 + \omega_0}\delta(\omega_1 - \omega_2). \tag{10.53}$$

So the Green function in frequency space reads:

$$G(\omega_n) = \frac{1}{i\omega_n - \omega_0} \tag{10.54}$$

with $\omega_n = \frac{2n\pi}{\beta}$.

10.1.3   Single Fermion

We now want to construct the path integral for a single fermion described by the hamiltonian:

$$H = \epsilon_0 c^\dagger c, \tag{10.55}$$

with:

$$\{c, c^\dagger\} = 1, \{c, c\} = \{c^\dagger, c^\dagger\} = 0 \tag{10.56}$$

In order to achieve this goal we need to introduce a basis of fermion coherent states as we did in the bosonic case. These are eigenstates of the fermion destruction operator:

$$c|\psi\rangle = \psi|\psi\rangle,$$
$$\langle\bar{\psi}|c^\dagger = \langle\bar{\psi}|\bar{\psi}. \tag{10.57}$$

where the coherent states for fermions are:

$$|\psi\rangle = e^{-\psi c^\dagger}|0\rangle,$$
$$\langle\bar{\psi}| = \langle 0|e^{-c\bar{\psi}}. \tag{10.58}$$

where, in contrast to bosons, $\psi, \bar{\psi}$ are Grassmann numbers. These anticommute despite of being numbers!

Algebra of Grassmann numbers

We can see why Grassmann numbers anticommute from:

$$cc|\psi\rangle = c^2|\psi\rangle = 0 = \psi^2|\psi\rangle \implies \psi^2 = 0 \tag{10.59}$$

Similarly, $\bar{\psi}^2 = 0$. We can summarize the Grassmann anticommutation relations as:

$$\{\psi, \bar{\psi}\} = \{\psi, \psi\} = \{\bar{\psi}, \bar{\psi}\} = 0. \tag{10.60}$$

Also Grassmann numbers anticommute with $c, c^\dagger$:

$$\{\psi, c\} = \{\psi, c^\dagger\} = 0 \tag{10.61}$$

$|\psi\rangle$ is an eigenstate of c:

$$|\psi\rangle = e^{-\psi c^\dagger}|0\rangle = (1 - \psi c^\dagger)|0\rangle, \tag{10.62}$$

since $\Psi^2 = 0$! Acting with c on $|\psi\rangle$:

$$c|\psi\rangle = c|0\rangle - c\psi c^\dagger|0\rangle = \psi(1 - c^\dagger c)|0\rangle = \psi|0\rangle. \tag{10.63}$$

Alternatively, since:

$$|0\rangle = \frac{|\psi\rangle}{1 - \psi c^\dagger} \tag{10.64}$$

we have:

$$c|\psi\rangle = \psi\left(\frac{1}{1 - \psi c^\dagger}\right)|\psi\rangle = \psi(1 + \psi c^\dagger)|\psi\rangle = \psi|\psi\rangle. \tag{10.65}$$

Grassmann numbers commute with bosonic observables:

$$[\psi, O] = [\bar{\psi}, O] = 0 \tag{10.66}$$

and anticommute with other fermion operators:

$$\{\psi, c\} = \{\bar{\psi}, c\} = \{\psi, c^\dagger\} = \{\bar{\psi}, c^\dagger\} = 0. \tag{10.67}$$

In the fermionic case, the Taylor expansion of any function of ψ, $\bar{\psi}$ truncates at first order since $\psi = \bar{\psi}^2 = 0$:

$$f(\bar{\psi}, \psi) = f_0 + \bar{f}_1\bar{\psi} + f_1\psi + f_{12}\bar{\psi}\psi. \tag{10.68}$$

The derivative of a function satisfies:

$$\frac{\partial}{\eta}(\bar{\eta}) = -\bar{\eta}\frac{\partial \eta}{\partial \eta} = -\bar{\eta} \tag{10.69}$$

Hence:

$$\frac{\partial f}{\partial \phi} = f_1 - f_{12}\bar{\phi}$$

$$\frac{\partial f}{\partial \bar{\phi}} = \bar{f}_1 + f_{12}\phi \tag{10.70}$$

The integration/derivation of Grassmann fermion fields satisfy:

$$\int d\psi 1 = \partial_\psi 1 = \int d\bar{\psi} 1 = 0$$

$$\int d\psi \psi = \int d\bar{\psi}\bar{\psi} = 1$$

$$\partial_\psi \psi = \partial_{\bar{\psi}}\bar{\psi} = 1$$

$$\int \psi d\psi = \int \bar{\psi}d\bar{\psi} = -1$$

$$\int d\bar{\psi}d\psi e^{-a\bar{\psi}\psi} = \int d\bar{\psi}d\psi(1 - a\psi\bar{\psi}) = a \tag{10.71}$$

The completeness relation

$$\langle \bar{\psi}|\psi'\rangle = \langle 0|(1 - c\bar{\psi}(1 - \psi'c^\dagger)|0\rangle = \langle 0|(1 - \psi'c^\dagger - c\bar{\psi} + c\bar{\psi}\psi'c^\dagger)|0\rangle = \langle 0|(1 + \bar{\psi}\psi')|0\rangle = e^{\bar{\psi}\psi'}$$

Hence, the identity read:

$$\int d\bar{\psi}d\psi e^{-\bar{\psi}\psi}|\psi\rangle\langle\bar{\psi}| = I \tag{10.72}$$

and the matrix elements:

$$\langle \bar{\psi}|A|\psi\rangle = A[\bar{\psi}, \psi]\langle\bar{\psi}|\psi\rangle = e^{\bar{\psi}\psi}A[\bar{\psi}, \psi].$$

Based on the above relation the trace of an operator is:

$$Tr A = \int d\bar{\psi}d\psi e^{-\bar{\psi}\psi}\langle -\bar{\psi}|A|\psi\rangle, \tag{10.73}$$

where the $-$ sign in $\langle -\psi|$ comes from:

$$\delta_{mn} = \langle n|m\rangle = \int d\bar{\psi}\,d\psi\, e^{-\bar{\psi}\psi}\langle -\bar{\psi}|m\rangle\langle n|\psi\rangle,$$

$$Tr\,A = \sum_{m,n}\langle m|A|n\rangle\delta_{nm} = \sum_{m,n}\int d\bar{\psi}\,d\psi\, e^{-\bar{\psi}\psi}\langle -\bar{\psi}|m\rangle\langle m|A|n\rangle\langle n|\Psi\rangle$$

$$= \int d\bar{\psi}\,d\psi\, e^{-\bar{\psi}\psi}\langle -\bar{\psi}|A|\psi\rangle. \tag{10.74}$$

Note that the integration measure for fermion coherent states is:

$$\int d\bar{\psi}\,d\psi\, e^{-\psi\bar{\psi}}, \tag{10.75}$$

which, compared to bosons containing the different normalization:

$$\int \frac{d\bar{\phi}\,d\phi}{-2\pi i} e^{-\psi\bar{\psi}}, \tag{10.76}$$

We finally conclude with the Gaussian integration including the source terms for fermions:

$$\int \prod_j d\bar{\psi}_j\,d\psi_j\, e^{-(\bar{\phi}A\phi-\bar{\eta}\psi-\bar{\psi}\eta)} = |A|e^{\bar{\eta}A^{-1}\eta}, \tag{10.77}$$

where $\bar{\eta},\eta$ are Grassmann numbers.

Partition function of a single fermion

We start with the hamiltonian of a single fermion:

$$H_0 = \epsilon_0 c^\dagger c. \tag{10.78}$$

The partition function can be obtained along the same line as for bosons using Grassmann numbers instead.

$$Z = Tr\{e^{-\beta H}\} = \sum_n \langle n|e^{-\beta H}|n\rangle = \sum_n \int d\bar{\psi}\,d\psi\, e^{-\bar{\psi}\psi}\langle n|\psi\rangle\langle\bar{\psi}|e^{-\beta H}|n\rangle$$

$$= \int d\bar{\psi}\,d\psi\, e^{-\bar{\psi}\psi}\langle -\bar{\psi}|e^{-\beta H}|\psi\rangle, \tag{10.79}$$

where $\langle -\bar{\psi}| \equiv \langle 0|e^{-\bar{\psi}c}$ since $\langle\bar{\psi}| \equiv \langle 0|e^{\bar{\psi}c}$. The scalar product is:

$$\langle -\bar{\psi}|\psi\rangle = \langle 0|e^{-\bar{\psi}c}e^{-\psi c^\dagger}|0\rangle \tag{10.80}$$

As in the bosonic case, we divide the temporal evolution operator into infinitesimal imaginary time steps:

$$\langle -\bar{\psi}| : (e^{-\epsilon H})^M : |\Psi\rangle. \tag{10.81}$$

and introducing it in the partition function:

$$Z = -\int d\bar{\psi}_M d\psi_0 e^{\bar{\psi}_M \psi_0} \prod_{j=1}^{M-1} d\bar{\psi}_j d\psi_j e^{-\bar{\psi}_j \psi_j} \prod_{j=1}^{M} \langle \bar{\psi}_j | e^{-\epsilon H} | \psi_{j-1} \rangle, \qquad (10.82)$$

now using the antiperiodicity property: $\psi_M = -\psi_0$, $\bar{\psi}_M = \bar{\psi}_0$ we have:

$$Z = \int \prod_{j=1}^{M} d\bar{\psi}_j d\psi_j e^{-\bar{\psi}_j \psi_j} \langle \bar{\Psi}_j | e^{-\epsilon H} | \psi_{j-1} \rangle, \qquad (10.83)$$

with: $\langle \bar{\psi}_j | e^{-\epsilon H} | \psi_{j-1} \rangle = e^{\bar{\psi}_j \psi_{j-1}} : e^{-H[\bar{\psi}_j, \psi_{j-1}]}$, or:

$$Z = lim_{M\to\infty} \int \prod_{j=1}^{M} d\bar{\psi}_j d\psi_j e^{-S} \qquad (10.84)$$

with:

$$S = \sum_{j=1}^{M} \left[\bar{\psi}_j \frac{(\psi_j - \psi_{j-1})}{\epsilon} + \epsilon_0 \bar{\psi}_j \psi_{j-1} \right] \epsilon \qquad (10.85)$$

which represents a sum over all possible histories of the fields: $\psi(\tau_j) = \{\psi_1, \psi_2, \ldots, \psi_M\}$, $\bar{\psi}(\tau_j) = \{\bar{\psi}_1, \bar{\psi}_2, \ldots, \bar{\psi}_M\}$, which in the $\epsilon \to 0$ limit becomes:

$$Z = \int_{\psi(0)=-\psi(\beta)} \mathcal{D}[\bar{\psi}(\tau)]\mathcal{D}[\psi(\tau)]e^{-\int_0^\beta d\tau\{\bar{\psi}(\tau)\frac{\partial}{\partial\tau}\psi(\tau)+H[\bar{\psi}(\tau),\psi(\tau)]\}} \qquad (10.86)$$

for fermions. For bosons the same expression holds with the periodic boundary condition, $\psi(0) = \psi(\beta)$ instead.

A representation in frequency space can be achieved by performing a Fourier transform of the fields:

$$\bar{\psi}(\tau) = \frac{1}{\sqrt{\beta}} \sum_n e^{i\omega_n \tau} \bar{\psi}(\omega_n)$$

$$\psi(\tau) = \frac{1}{\sqrt{\beta}} \sum_n e^{-i\omega_n \tau} \psi(\omega_n)$$

$$\frac{\partial\psi(\tau)}{\partial\tau} = \frac{1}{\sqrt{\beta}} \sum_n (-i\omega_n)e^{-i\omega_n \tau} \psi(\omega_n) \qquad (10.87)$$

where $\omega_n = \frac{(2n+1)\pi}{\beta}$ (fermions), $\omega_n = \frac{2n\pi}{\beta}$ (bosons). The frequency decomposition above is obtained from the antiperiodicity (periodicity) for fermions (bosons).

In terms of the Fourier components the partition function reads:

$$Z = \int \mathcal{D}\bar{\psi}\mathcal{D}\psi\, e^{-\frac{1}{\beta}\sum_{m,n}\int_0^\beta d\tau\, e^{i(\omega_n-\omega_m)\tau}(-i\omega_m+\epsilon_0)\bar{\psi}(\omega_n)\psi(\omega_m)}$$

$$= \int \mathcal{D}\bar{\psi}\mathcal{D}\psi\, e^{-\sum_n \bar{\psi}(\omega_n)(-i\omega_n+\epsilon_0)\psi(\omega_n)} \qquad (10.88)$$

Performing the Gaussian integral:

$$Z = \prod_n (-i\omega_n + \epsilon_0). \qquad (10.89)$$

The path integral representation of Z in the $T \to 0$ ($\beta \to \infty$) limit reduces to:

$$lim_{\beta\to\infty}\frac{1}{\beta}\sum_n \to \frac{1}{\beta}\int \frac{\delta\omega_n}{(\frac{2\pi}{\beta})} \equiv \int \frac{d\omega}{2\pi} \qquad (10.90)$$

Thus:

$$Z = \int \mathcal{D}\bar{\psi}(\omega)\mathcal{D}\psi(\omega)e^{\int \frac{d\omega}{2\pi}\bar{\psi}(\omega)(i\omega-\epsilon_0)\psi(\omega)} \qquad (10.91)$$

in the $T \to 0$ limit.

Green's function for free fermions

We shall include source terms in the original hamiltonian by writing:

$$H = \epsilon_0 c^\dagger c - \bar{\eta}(\tau)c - c^\dagger \eta(\tau) \qquad (10.92)$$

The partition function in the presence of source terms becomes the generating functional:

$$Z[\bar{\eta}, \eta] = \int \mathcal{D}\bar{\psi}\mathcal{D}\psi\, e^{-S+\int_0^\beta [\bar{\eta}(\tau)\psi(\tau)+\bar{\psi}(\tau)\eta(\tau)]}, \qquad (10.93)$$

where the action reads:

$$S = \int_0^\beta d\tau(\bar{\psi}(\tau)(\partial_\tau + \epsilon_0)\psi(\tau)) \qquad (10.94)$$

The time-ordered products or Green's functions can be obtained from differentiation of the generating functional:

$$G(\tau - \tau') = -\frac{\delta^2 \ln Z}{\delta\eta(\tau')\delta\bar{\eta}(\tau)}\Big|_{\bar{\eta},\eta=0} = -(\partial_\tau + \epsilon_0)^{-1} = -\langle \psi(\tau)\bar{\psi}(\tau')\rangle \qquad (10.95)$$

Hence, we can associate the inverse of the Gaussian coefficient in the action with the imaginary-time Green's function of the non-interacting fermions:

$$S = \int_0^\beta \bar{\psi}(\tau)(-G_0^{-1}(\tau))\psi(\tau). \tag{10.96}$$

The free energy of a free fermion can be obtained automatically from:

$$F = -k_B T \ln Z = -k_B T \ln |\partial_\tau + \hat{h}| = -k_B T Tr \ln(-G_0^{-1}). \tag{10.97}$$

The Green function can be expressed in Matsubara frequencies by using the Fourier decomposition of the Grassmann fields, $\phi(\tau)$. In this basis:

$$\partial_\tau + \epsilon_0 \to -i\omega_n + \epsilon_0$$
$$G^{-1} = (-i\omega_n + \epsilon_0). \tag{10.98}$$

Hence:

$$G(\omega_n) = \frac{1}{i\omega_n - \epsilon_0}, \tag{10.99}$$

and the action reads:

$$S = \sum_n (-i\omega_n + \epsilon_0)\bar{\phi}(\omega_n)\phi(\omega_n) = \sum_n (-G_0^{-1}(i\omega_n))\bar{\phi}(\omega_n)\phi(\omega_n). \tag{10.100}$$

The generating functional can be obtained by evaluating the Gaussian integrals:

$$Z[\bar{\eta}, \eta] = \prod_n (-i\omega_n + \epsilon_0)e^{\sum_n \bar{\eta}(\omega_n)(-i\omega_n+\epsilon_0)^{-1}\eta(\omega_n)}. \tag{10.101}$$

By differentiating:

$$G(\omega_n, \omega_{n'}) = -\langle \psi(\omega_n)\bar{\psi}(\omega_{n'})\rangle = -\frac{\delta^2 \ln Z}{\delta\eta(\omega_{n'})\delta\bar{\eta}(\omega_n)} = \frac{\delta(\omega_n - \omega_{n'})}{i\omega_n - \epsilon_0} \tag{10.102}$$

So the Green function in Matsubara frequencies of a free single fermions is:

$$G(\omega_n, \omega_{n'}) = \frac{\delta(\omega_n - \omega_{n'})}{i\omega_n - \epsilon_0} \tag{10.103}$$

In a similar way, the free energy in Matsubara frequencies reads:

$$F = -k_B T \sum_n \ln(-i\omega_n + \epsilon_0). \tag{10.104}$$

Performing the Matsubara frequencies one recovers the well-known free energy expression of a single fermion: $F = -k_B \ln(1 + e^{-\beta\epsilon})$.

10.1.4   *Wick's Theorem and Perturbation Theory*

Based on the path integral formalism we may also derive Wick's theorem in a straightforward way. Let's assume the hamiltonian:

$$H = H_0 + V, \tag{10.105}$$

where H_0 is the non-interacting part:

$$H_0 + \sum_{\mathbf{k}\sigma} \epsilon_{\mathbf{k}} c^{\dagger}_{\mathbf{k}\sigma} c_{\mathbf{k}\sigma}, \tag{10.106}$$

and V the Coulomb interaction:

$$V = \frac{1}{V} \sum_{\mathbf{k},\mathbf{k}',\mathbf{q},\sigma,\sigma'} \frac{V(\mathbf{k}-\mathbf{k}')}{2} c^{\dagger}_{\mathbf{k}'+\mathbf{q},\sigma'} c^{\dagger}_{\mathbf{k}-\mathbf{q},\sigma} c_{\mathbf{k}\sigma} c_{\mathbf{k}'\sigma'}. \tag{10.107}$$

The Green function for a fermion can be expressed in terms of the non-interacting part of the action as:

$$\begin{aligned}
G_{\alpha\beta}(\mathbf{k}, \tau - \tau') &= -\frac{1}{Z} \int \mathcal{D}\bar{\psi}\mathcal{D}\psi\, \psi_{\alpha}(\mathbf{k},\tau)\bar{\psi}_{\beta}(\mathbf{k},\tau') e^{-S} \\
&= -\frac{1}{Z}\frac{Z_0}{Z_0} \int \mathcal{D}\bar{\psi}\mathcal{D}\psi\, e^{-S_0} e^{-\int_0^\beta d\tau V[\bar{\psi},\psi]} \\
&= -\frac{\langle \psi_{\alpha}(\mathbf{k},\tau)\bar{\psi}_{\beta}(\mathbf{k},\tau') e^{-S_V}\rangle_0}{(Z/Z_0)} = -\frac{\langle \psi_{\alpha}(\mathbf{k},\tau)\bar{\psi}_{\beta}(\mathbf{k},\tau') e^{-S_V}\rangle_0}{\langle e^{-S_V}\rangle_0},
\end{aligned}$$

where, in the last step, we have used the relation:

$$\frac{Z}{Z_0} = \frac{1}{Z_0}\int \mathcal{D}\bar{\psi}\mathcal{D}\psi\, e^{-S_V} e^{-S_0} = \langle e^{-S_V}\rangle_0. \tag{10.108}$$

When we expand e^{-S_V} in a Taylor series we will encounter quantities like:

$$\langle \psi_1 \psi_2 \ldots \bar{\psi}_1 \bar{\psi}_2 \ldots\rangle_0. \tag{10.109}$$

which can be evaluated using Wick's theorem.

Proof of Wick's theorem with path integrals

Let's assume that:

$$S_0 = \bar{\Psi} A \Psi, \tag{10.110}$$

so that:

$$\langle \Psi_1 \Psi_2 \ldots \bar{\Psi}_{1'} \bar{\Psi}_{2'} \ldots \rangle_0 = \frac{1}{|A|} \int \mathcal{D}\bar{\Psi} \mathcal{D}\Psi \, e^{-\bar{\Psi} A \Psi} \Psi_1 \Psi_2 \ldots \bar{\Psi}_{1'} \bar{\Psi}_{2'} \ldots \quad (10.111)$$

We now use the identity:

$$Z[\bar{\eta}, \eta] = \int \mathcal{D}\bar{\Psi} \mathcal{D}\Psi \, e^{-\bar{\Psi} A \Psi + \bar{\eta}\Psi + \bar{\Psi}\eta} = |A| e^{\bar{\eta} A^{-1} \eta}, \quad (10.112)$$

to generate the average value, $\langle\rangle_0$ above by just taking derivatives at both sides of the identity:

$$\frac{\partial^{2n} Z[\bar{\eta}, \eta]}{\partial \bar{\eta}_1 \partial \bar{\eta}_2 \ldots \partial \eta_{1'} \eta_{2'} \ldots}\Big|_{\eta=\bar{\eta}=0} = (-1)^n \int \mathcal{D}\bar{\Psi} \mathcal{D}\Psi \, e^{-\bar{\Psi} A \Psi} \Psi_1 \Psi_2 \ldots \bar{\Psi}_{1'} \bar{\Psi}_{2'}$$

$$= (-1)^n |A| \langle \Psi_1 \Psi_2 \ldots \bar{\Psi}_{1'} \bar{\Psi}_{2'} \ldots \rangle_0. \quad (10.113)$$

Now we take the derivative on the r.h.s of the identity above:

$$\frac{\partial^{2n} |A| e^{\bar{\eta} A^{-1} \eta}}{\partial \bar{\eta}_1 \partial \bar{\eta}_2 \ldots \partial \eta_{1'} \partial \eta_{2'} \ldots}\Big|_{\eta=\bar{\eta}=0} = |A|(-1)^n \frac{\partial}{\partial \bar{\eta}_1 \partial \bar{\eta}_2 \ldots} \left(\sum_{k_n} \bar{\eta}_{k_n} A_{k_n n'}^{-1} \right) \ldots$$

$$\left(\sum_{k_1} \bar{\eta}_{k_1} A_{k_1 1'}^{-1} \right) e^{\bar{\eta} A^{-1} \eta}\Big|_{\eta=\bar{\eta}=0} = (-1)^n |A| \sum_P (-1)^P A_{P_n n'}^{-1} A_{P_{n-1} n - 1'}^{-1} \ldots A_{P_2 2'}^{-1} A_{P_1 1'}^{-1}. \quad (10.114)$$

Since $A_{ij}^{-1} = G_{ij}^0$, this is a proof of Wick's theorem:

$$\langle \Psi_1 \Psi_2 \ldots \bar{\Psi}_{1'} \bar{\Psi}_{2'} \ldots \rangle_0 = \frac{1}{|A|} \int \mathcal{D}\bar{\Psi} \mathcal{D}\Psi \, e^{-\bar{\Psi} A \Psi} \Psi_1 \Psi_2 \ldots \bar{\Psi}_{1'} \bar{\Psi}_{2'}$$

$$= \sum_P (-1)^P A_{P_n n}^{-1} \ldots A_{P_2 2'}^{-1} A_{P_1 1'}^{-1}. \quad (10.115)$$

Now we have all the ingredients for applying FDPT. Diagrams are generated by expanding e^{-S_V} in the numerator and denominator of the Green's functions:

$$G_{\alpha\beta} = -\frac{\langle \Psi_\alpha \bar{\Psi}_\beta e^{-S_V} \rangle_0}{\langle e^{-S_V} \rangle_0} \quad (10.116)$$

Note that e^{-S_V} can be expanded as a normal function through the Taylor series since the argument of the exponential is a number:

$$e^{-S_V} = \sum_n \frac{(-1)^n}{n!} \left[\int_0^\beta d\tau\, V[\bar{\Psi}, \Psi] \right]^n, \qquad (10.117)$$

where normal ordering in $V[\bar{\Psi}, \Psi]$ is assumed. This can be expressed as sum over disconnected Feynman diagrams.

On the other hand the Taylor expansion of the numerator in Eq. (10.116) can be factorized as the product of a connected and a disconnected sum of Feynman diagrams. Since the disconnected part cancels the denominator, only connected Feynman diagrams enter the expansion:

$$G_{\alpha\beta} = -\langle \Psi_\alpha \bar{\Psi}_\beta e^{-S_V} \rangle_{0,conn.} \qquad (10.118)$$

Exercise III.1: Wick's theorem from path integration

Wick's theorem can be demonstrated straightforwardly using path integration. The two particle time-ordered product is defined as:

$$\langle \Psi_1 \Psi_2 \bar{\Psi}_{1'} \bar{\Psi}_{2'} \rangle_0 = \frac{1}{Z} \int D\bar{\Psi} D\Psi\, \Psi_1 \Psi_2 \bar{\Psi}_{1'} \bar{\Psi}_{2'} e^{-S_0},$$

with the Gaussian action: $S_0 = \int_0^\beta d\tau\, \bar{\Psi}(\tau)(\partial_\tau + h)\Psi(\tau)$, and $Z = \int D\bar{\Psi} D\Psi e^{-S_0}$.
Using the generating functional:

$$Z[\bar{\eta}, \eta] = \int D\bar{\Psi} D\Psi e^{-\bar{\Psi}A\Psi + \bar{\eta}\Psi + \bar{\Psi}\eta} = |A| e^{\bar{\eta}A^{-1}\eta},$$

(where: $A = (\partial_\tau + h)$, is the matrix defining the Gaussian action) demonstrate that the two particle Green function can be decomposed in terms of a sum of products of two single particle Green functions as:

$$\langle \Psi_1 \Psi_2 \bar{\Psi}_{1'} \bar{\Psi}_{2'} \rangle_0 = -G_{11'} G_{22'} + G_{12'} G_{21'}$$

since: $G_{ij'} = A_{ij'}^{-1}$.

This exercise is set and solved in [6].

Exercise III.2: Electron-phonon coupling

Electrons in solids are affected by the vibrations of the host ions in the lattice, the phonons. This coupling mechanism generates a net attractive interaction between the electrons which is ultimately responsible for Cooper pairing and is key to conventional BCS superconductivity. Starting from the electron-phonon hamiltonian:

$$H = H_{el} + H_{ph} + H_{el-ph}$$

$$= \sum_{k\sigma} \epsilon_k c_{k\sigma}^\dagger c_{k\sigma} + \sum_{q\sigma} \omega_{q\lambda} \left(b_{q\lambda}^\dagger b_{q\lambda} + \frac{1}{2} \right)$$

$$+ \sum_{k,q,\lambda} g_{q\lambda} c_{k+q\sigma}^\dagger c_{k\sigma} [b_{q\lambda} + b_{-q\lambda}^\dagger],$$

(a) Formulate the coherent state action of the electron-phonon system introducing Grassmann fields to describe the electron operators and complex fields for the phonon operators. Use the Matsubara frequency representation for representing the fermion and phonon fields.

(b) Integrate out the phonon fields in the functional integral (Gaussian integration of complex fields) and write down the effective action for the electrons that arises. Show that an attractive interaction between the electrons can be effectively generated by the lattice vibrations. Compare the effective potential obtained in this way with the one discussed based on the operator formalism.

This exercise is set and solved in [10].

10.2 Path Integral Formulation for Non-equilibrium Systems

As discussed in Chap. 7, the concept of a partition function can be extended to non-equilibrium situations. In fact, mean values of an observable $\mathcal{O}$ can be obtained from a non-equilibrium partition function $Z[\chi]$ defined as

$$Z[\chi] = \text{Tr}\left[\mathcal{U}_c[\chi] \rho(-\infty) \right] / \text{Tr}[\rho],$$

where $\mathcal{U}_c[\chi]$ denotes the evolution operator over the Keldysh contour in the presence of a source term $\chi(t)\mathcal{O}$ added to the system Hamiltonian. As in the equilibrium case, a first step in the path integral formulation of the theory is to obtain a functional integral representation of this partition function.

As a first simple example let us consider the case of single bosonic quantum level model, $H = \omega_0 b^\dagger b$ as in Sect. 10.1. To better understand the Keldysh structure in the calculation of Z let us consider a discretization of the contour as the one shown

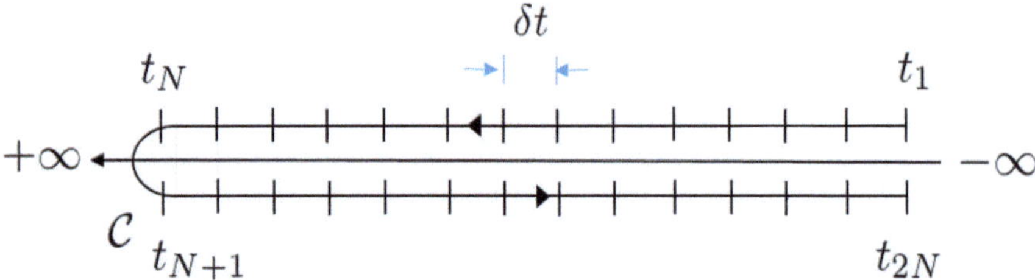

Fig. 10.1 Discretized Keldysh contour to set up the functional integral representation of the partition function

in Fig. 10.1. Then we can write $\mathcal{U}_\mathcal{C}$ as

$$\mathcal{U}_\mathcal{C} \simeq e^{iH(t_{2N-1})\delta t} \dots e^{iH(t_{N+1})\delta t}\mathbb{I}e^{-iH(t_{N-1})\delta t} \dots e^{-iH(t_1)\delta t} , \tag{10.119}$$

where the identity $\mathbb{I}$ between t_N and t_{N+1} indicates that there is no evolution between these two times as they correspond to the same physical instant.

We next introduce a representation of the identity in terms of coherent states $|\phi_j\rangle$ associated to each instant t_j, i.e.,

$$\mathbb{I} = \int d[\bar{\phi}_j, \phi_j]e^{-|\phi_j|^2}|\phi_j\rangle\langle\phi_j| , \tag{10.120}$$

thus

$$\mathrm{Tr}[\mathcal{U}_\mathcal{C}\hat{\rho}_0] \simeq \int \prod_j d[\bar{\phi}_j, \phi_j]e^{-\sum_j |\phi_j|^2} \langle\phi_{2N}|e^{iH\delta t}|\phi_{2N-1}\rangle \dots \langle\phi_{N+2}|e^{iH\delta t}|\phi_{N+1}\rangle$$

$$\times \langle\phi_{N+1}|\mathbb{I}|\phi_N\rangle\langle\phi_N|e^{-iH\delta t}|\phi_{N-1}\rangle \dots \langle\phi_2|e^{-iH\delta t}|\phi_1\rangle\langle\phi_1|\hat{\rho}_0|\phi_{2N}\rangle \tag{10.121}$$

where $\hat{\rho}_0 = e^{-\beta(H-\mu N)}$ is the initial density matrix, assumed to be the equilibrium one. For the next step we need the following relations

$$\langle\phi_j|e^{\pm iH\delta t}|\phi_{j-1}\rangle \simeq e^{\bar{\phi}_j\phi_{j-1}}e^{h_\pm\bar{\phi}_j\phi_{j-1}}$$

$$\langle\phi_1|\hat{\rho}_0|\phi_{2N}\rangle = e^{\bar{\phi}_1\phi_{2N}\rho(\omega_0)}$$

$$\langle\phi_{N+1}|\mathbb{I}|\phi_N\rangle = e^{\bar{\phi}_{N+1}\phi_N} , \tag{10.122}$$

where $h_\pm = 1 \pm i\omega_0\delta t$ and $\rho(\omega_0) = e^{-\beta(\omega_0-\mu)}$. Based on these relations one can derive the expression for the partition function as a functional integral

$$Z = \frac{1}{\mathrm{Tr}\hat{\rho}_0} \int \prod_{j=1}^{2N} d[\bar{\phi}_j, \phi_j] \exp\left(i \sum_{j,j'=1}^{2N} \bar{\phi}_j G_{jj'}^{-1}\phi_j\right) \tag{10.123}$$

where

$$
i\hat{G}^{-1} \equiv \left(
\begin{array}{cccc|cccc}
-1 & & & & & & & \rho(\omega_0) \\
h_- & -1 & & & & & & \\
 & \ddots & \ddots & & & & & \\
 & & h_- & -1 & & & & \\
\hline
 & & & 1 & -1 & & & \\
 & & & & h_+ & -1 & & \\
 & & & & & \ddots & \ddots & \\
 & & & & & & h_+ & -1
\end{array}
\right) .
$$

(10.124)

As can be observed, $\hat{G}^{-1}$ has a simple lower bi-diagonal structure, except for the $(1, 2N)$ and $(N+1, N)$ elements coming from the initial condition and the contour closing respectively. To make this representation useful we can add source terms so that

$$
Z[\bar{J}, J] = \frac{1}{\text{Tr}\hat{\rho}_0} \int \prod_{j=1}^{2N} d[\bar{\phi}_j, \phi_j] \exp\left[i \sum_{j,j'=1}^{2N} \bar{\phi}_j G_{jj'}^{-1} \phi_j + \sum_{j=1}^{2N} \left(\bar{\phi}_j J_j + \phi_j \bar{J}_j\right) \right]
$$

$$
= \frac{1}{\text{Tr}\hat{\rho}_0} \frac{\exp\left[i \sum_{jj'} J_j G_{jj'} \bar{J}_{j'} \right]}{\det\left[-i\hat{G}^{-1} \right]} ,
$$

(10.125)

where in the last step we have used the known result for the bosonic Gaussian integral. From Eq. (10.125) one can readily show that

$$
\langle \phi_j \bar{\phi}'_j \rangle = \frac{\delta^2 Z}{\delta \bar{J}_j \delta J_{j'}} \Big\rfloor_{J,\bar{J}=0} = i G_{jj'}
$$

(10.126)

In order to obtain explicit expressions for $\hat{G}$ it is convenient to notice the block structure of G^{-1} indicated by vertical and horizontal lines in (10.124), which allows us to decompose it as

$$
\hat{G} = \begin{pmatrix} \hat{G}^{++} & \hat{G}^{+-} \\ \hat{G}^{-+} & \hat{G}^{--} \end{pmatrix} .
$$

One can then show that (Exercise III.3)

$$G_{jj'}^{++} = -ih_-^{j-j'} \left(\theta_{jj'} + \frac{\rho(\omega_0)(h_-h_+)^{N-1}}{1 - \rho(\omega_0)(h_-h_+)^{N-1}} \right)$$

$$G_{jj'}^{--} = -ih_+^{j-j'} \left(\theta_{jj'} + \frac{\rho(\omega_0)(h_-h_+)^{N-1}}{1 - \rho(\omega_0)(h_-h_+)^{N-1}} \right)$$

$$G_{jj'}^{+-} = -i\frac{\rho(\omega_0)h_+^{N-j'} h_-^{j-1}}{1 - \rho(\omega_0)(h_-h_+)^{N-1}}$$

$$G_{jj'}^{-+} = -i\frac{h_-^{N-j'} h_+^{j-1}}{1 - \rho(\omega_0)(h_-h_+)^{N-1}} \;, \tag{10.127}$$

where $\theta_{ij} = 1$ for $i - j \geq 0$ and zero otherwise. It is now straightforward to the reach the continuous limit. For that we assign $t \equiv j\delta t$ on the $+$ branch, $t \equiv (N - 1)\delta t$ on the $-$ branch, $h_-^j \to e^{-i\omega_0 t}$ and $h_+^{N-j} \to e^{i\omega_0 t}$, which leads to

$$G^{++}(t, t') = -ie^{-i\omega_0(t-t')} \left(\theta(t - t') + n_B(\omega_0) \right)$$

$$G^{--}(t, t') = -ie^{-i\omega_0(t-t')} \left(\theta(t' - t) + n_B(\omega_0) \right)$$

$$G^{+-}(t, t') = -ie^{-i\omega_0(t-t')} n_B(\omega_0)$$

$$G^{-+}(t, t') = -ie^{-i\omega_0(t-t')} \left(n_B(\omega_0) + 1 \right) \;, \tag{10.128}$$

where we have used $n_B(\omega_0) = \rho(\omega_0)/(1 - \rho(\omega_0))$. As can be checked, these results coincide with the ones obtained in Exercise II.1.

It is also instructive to analyze the continuous limit in the functional integral representation of Z. To this end we can express Z as

$$Z = \int \mathcal{D}[\bar{\phi}, \phi] \exp i S[\bar{\phi}, \phi] \tag{10.129}$$

where the action $S[\bar{\phi}, \phi]$ corresponds to $\sum_{jj'} \bar{\phi}_j G_{jj'}^{-1} \phi_j$. Using the expression for $G_{jj'}^{-1}$ we get

$$S[\bar{\phi}, \phi] = \sum_{j=2}^{2N} \delta t_j \left[i\bar{\phi}_j \left(\frac{\bar{\phi}_j - \phi_{j-1}}{\delta t_j} \right) - \omega_0 \bar{\phi}_j \phi_{j-1} \right] + i\bar{\phi}_1 \left(\phi_1 - i\rho(\omega_0)\phi_{2N} \right) - \delta t \omega_0 \bar{\phi}_{N+1} \phi_N \tag{10.130}$$

where $\delta t_j = \delta t$ for $j \leq N$ and $-\delta t$ for $j > N$. Now, if we take the limit $\delta t \to 0$, $N \to \infty$ we can express

$$S[\bar{\phi}, \phi] = \int_{\mathcal{C}} dt \bar{\phi}(t) (i\partial_t - \omega_0) \phi(t)$$

$$= \int_{-\infty}^{\infty} dt \left[\bar{\phi}^+(t) (i\partial_t - \omega_0) \phi^+(t) - \bar{\phi}^-(t) (i\partial_t - \omega_0) \phi^-(t) \right] \;. \tag{10.131}$$

As one can observe, in Eq. (10.131) we have lost track of the boundary conditions at the beginning and at the closing of the contour, and thus the two parts of the action appear to be disconnected. However, it is necessary to remember that $(i\partial_t - \omega_0)$ stands for G^{-1} and the GFs *do know* about the bondary conditions. To make this more explicit let us write the action in a full Keldysh (matrix) form, i.e.,

$$S[\bar{\phi}, \phi] = \int_{-\infty}^{\infty} dt \, (\bar{\phi}^+ \; \bar{\phi}^-) \begin{pmatrix} i\partial_t - \omega_0 & 0 \\ 0 & -(i\partial_t - \omega_0) \end{pmatrix} \begin{pmatrix} \phi^+ \\ \phi^- \end{pmatrix} . \tag{10.132}$$

If we now perform the Keldysh rotation $\hat{U} = (1 - i\sigma_y)/\sqrt{2}$ introduced in Sect. 8.5 we obtain

$$S[\bar{\phi}, \phi] = \int_{-\infty}^{\infty} dt \, (\bar{\phi}_q \; \bar{\phi}_c) \begin{pmatrix} 0 & i\partial_t - \omega_0 \\ i\partial_t - \omega_0 & 0 \end{pmatrix} \begin{pmatrix} \phi_q \\ \phi_c \end{pmatrix} , \tag{10.133}$$

where $\phi_{c,q} = (\phi^+ \pm \phi^-)/\sqrt{2}$ are denoted, respectively, the *classical* and *quantum* components of the field. In frequency representation we would have

$$S[\bar{\phi}, \phi] = \int \frac{d\omega}{2\pi} \hat{\bar{\phi}}_\omega \hat{G}^{-1}(\omega) \hat{\phi}_\omega , \tag{10.134}$$

where $\hat{\phi}_\omega = (\phi_q(\omega) \; \phi_c(\omega))^T$ and

$$\hat{G}^{-1}(\omega) = \begin{pmatrix} 0 & G^A(\omega) \\ G^R(\omega) & G^K(\omega) \end{pmatrix}^{-1} = \begin{pmatrix} \left(\hat{G}^{-1}\right)^K & 1/G^R(\omega) \\ 1/G^A(\omega) & 0 \end{pmatrix} .$$

In this expression we should notice that $\left(G^{-1}\right)^K = -(1/G^R)G^K(1/G^A) \neq 1/G^K$. By remembering the results for the single quantum level $G^{R,A}(\omega) = 1/(\omega - \omega_0 \pm i0^+)$ and $G^K(\omega) = F(\omega)(G^R(\omega) - G^A(\omega))$ we finally get

$$S[\bar{\phi}, \phi] = \int \frac{d\omega}{2\pi} (\bar{\phi}_q(\omega) \; \bar{\phi}_c(\omega)) \begin{pmatrix} 2i0^+ F(\omega) & \omega - \omega_0 + i0^+ \\ \omega - \omega_0 - i0^+ & 0 \end{pmatrix} \begin{pmatrix} \phi_q(\omega) \\ \phi_c(\omega) \end{pmatrix} ,$$

where the infinitesimal imaginary part $i0^+$ allows for the convergence of the Fourier transform and puts in evidence the boundary conditions which were hidden in the time representation of the action. Such trick is not necessary when considering open quantum systems coupled to reservoirs.

Following a similar procedure one can obtain the path integral representation of the non-equilibrium partition function for a fermionic system. Again we can consider as an example a minimal model of a single quantum level, $H = \epsilon c^\dagger c$ and introduce coherent fermionic states $|\Psi\rangle$ such that $c|\psi\rangle = \psi|\psi\rangle$. Then, repeating the arguments used for the bosonic case but using the corresponding properties for Grassman variables one obtains

$$Z = \frac{1}{\mathrm{Tr}\hat{\rho}_0} \int \prod_{j=1}^{2N} d[\bar{\psi}_j, \psi_j] \exp\left[i \sum_{j,j'=1}^{2N} \bar{\psi}_j G_{jj'}^{-1} \psi_j \right] \tag{10.135}$$

where

$$i\hat{G}^{-1} \equiv \left(\begin{array}{ccccc|cccc} -1 & & & & & & & & -\rho(\epsilon) \\ h_- & -1 & & & & & & & \\ & \ddots & \ddots & & & & & & \\ & & h_- & -1 & & & & & \\ \hline & & & 1 & -1 & & & & \\ & & & & h_+ & -1 & & & \\ & & & & & \ddots & \ddots & & \\ & & & & & & h_+ & -1 \end{array}\right). \tag{10.136}$$

and $h_\pm = 1 \pm i\epsilon\delta t$. As can be observed, the only change in $\hat{G}^{-1}$ with respect to the bosonic case is the change of sign in $\rho(\epsilon)$ in the $(1, 2N)$ component. On the other hand, the fermionic Gaussian integrals are somewhat different to the bosonic ones. For these we have

$$\begin{aligned} Z[\bar{\chi}, \chi] &= \frac{1}{\mathrm{Tr}\hat{\rho}_0} \int \prod_{j=1}^{2N} d[\bar{\psi}_j, \psi_j] \exp\left[i \sum_{j,j'=1}^{2N} \bar{\psi}_j G_{jj'}^{-1} \psi_j + \sum_{j=1}^{2N} \left(\bar{\psi}_j \chi_j + \psi_j \bar{\chi}_j \right) \right] \\ &= \frac{\det\left[-i\hat{G}^{-1} \right]}{\mathrm{Tr}\hat{\rho}_0} \exp\left[i \sum_{jj'} \chi_j G_{jj'} \bar{\chi}_{j'} \right]. \end{aligned} \tag{10.137}$$

Using these properties one can show that

$$\langle \psi_j \bar{\psi}_{j'} \rangle = \frac{\delta^2 Z}{\delta\chi_j \delta\bar{\chi}_{j'}} \rfloor_{\chi=0} = iG_{jj'}, \tag{10.138}$$

and the normalization condition for Z is

$$Z[0] = \frac{\det\left[-i\hat{G}^{-1} \right]}{\mathrm{Tr}\hat{\rho}_0} = 1 \tag{10.139}$$

which implies $\det\left[-i\hat{G}^{-1} \right] = \mathrm{Tr}\hat{\rho}_0 = 1 + \rho(\epsilon)$.

Repeating the calculation done for the bosonic case and taking into account the appropriate changes, in the continuous limit one obtains

$$G^{+-}(t, t') = i n_F(\epsilon) e^{-i\epsilon(t-t')}$$

$$G^{-+}(t, t') = i \left(n_F(\epsilon) - 1 \right) e^{-i\epsilon(t-t')}$$

$$G^{++}(t, t') = \theta(t - t') G^{-+}(t, t') + \theta(t' - t) G^{+-}(t, t')$$

$$G^{--}(t, t') = \theta(t - t') G^{+-}(t, t') + \theta(t' - t) G^{-+}(t, t') \, , \tag{10.140}$$

with $n_F(\epsilon) = 1/(\rho(\epsilon) + 1)$, which again recovers the results obtained in Exercise II.1 for the corresponding case.

10.2.1 The Case of Lattice Models for Electron Transport

In Chap. 9 we discussed the application of the non-equilibrium formalism to transport in nanostructures based on lattice models. We here sketch how this task can be approached from the point of view of non-equilibrium path integral methods. Let us stress that the results obtained from both approaches should be equivalent and it is a matter of taste whether to follow one or the other.

As in Chap. 9 let us consider a two terminal normal system which can be described by a Hamiltonian $H = H_L + H_R + V$ which correspond to lattice models of the type $H_\mu = \sum_{j,j' \in \mu} h_{i,j'} c_j^\dagger c_{j'}$ for $\mu \equiv L, R$ and $V = V' \left(c_L^\dagger c_R + \text{h.c.} \right)$. To formulate the problem in field theoretical form we associate fermionic fields ψ_j to each lattice site so that the corresponding action $S = S_L + S_R + S_V$ can be written as

$$S_{L,R} = \int_{\mathcal{C}} dt \sum_{j,j' \in L,R} \bar{\psi}_j \left(i\delta_{ij'} \partial_t - h_{j,j'} \right) \psi_{j'}$$

$$S_V = -\int_{\mathcal{C}} dt \left(\bar{\psi}_L \ \bar{\psi}_R \right) \begin{pmatrix} 0 & V' \\ V' & 0 \end{pmatrix} \begin{pmatrix} \psi_L \\ \psi_R \end{pmatrix} . \tag{10.141}$$

We can now perform the usual steps within the non-equilibrium formalism:

* go to the "full Keldysh" representations, i.e., replace $\int_{\mathcal{C}} dt$ by $\int_{-\infty}^{\infty} dt$ and ψ_j by $\hat{\psi}_j = \left(\psi_j^+ \psi_j^- \right)^T$.

* shift to frequency representation, i.e., $\int dt \to \int \frac{d\omega}{2\pi}$ and $\hat{\psi}_j(t) \to \hat{\psi}_{j\omega}$.

* Integrate out the fields inside the leads, i.e., all $\hat{\psi}_j$ with $j \neq L, R$.

We thus obtain the following effective action

$$S_{\text{eff}} = \int \frac{d\omega}{2\pi} \left(\hat{\bar{\psi}}_{L\omega} \ \hat{\bar{\psi}}_{R\omega} \right) \begin{pmatrix} \hat{g}_L^{-1}(\omega) & -\hat{V}' \\ -\hat{V}' & \hat{g}_R^{-1}(\omega) \end{pmatrix} \begin{pmatrix} \hat{\psi}_{L\omega} \\ \hat{\psi}_{R\omega} \end{pmatrix} , \tag{10.142}$$

where the $\hat{\ }$ symbol corresponds to matrix structure in Keldysh space. One can then identify the coupled system GFs as

$$\check{\hat{G}}^{-1}(\omega) = \begin{pmatrix} \hat{g}_L^{-1}(\omega) & -\hat{V}' \\ -\hat{V}' & \hat{g}_R^{-1}(\omega) \end{pmatrix},$$

(10.143)

where the $\check{}$ symbol indicates matrix structure in the $L - R$ space. Once this identification has been established the transport properties can be analyzed as described in Chap. 9.

Chapter 11
Application of Path Integral Methods: The Renormalization Group Approach

The renormalization group (RG) method is a powerful and efficient tool to explore interacting fermions. Initially it was developed in the context of QFT to cure infinities arising in QED. Wilson and Kogut (1974), Wilson (1975), based on Kadanoff et al. (1967) approach to analyze critical exponents and critical phenomena in classical statistical mechanics, developed a scheme to tackle with systems in which fluctuations at all energy scales can occur. Wilson's approach had and still has a tremendous impact in condensed matter physics, particle physics and general statistical mechanics. Actually, Kadanoff and Wilson's work showed how the RG is a much broader approach than initially thought as a method to deal with the infinities arising in QED.

The essential idea underlying Wilson's RG method is the integration over fast or high energy modes of a model in order to generate an effective model describing the slow low energy degrees of freedom. In real space, this means that short range fluctuations are averaged out in favour of the long range fluctuations.

The method requires that the original initial model can be written in terms of a path integral. The important underlying concepts in the RG are: scale invariance, fixed points and universality. We will illustrate these ideas applying the RG first to a free fermion model. Subsequently we will illustrate the full power of the RG in the 1D Hubbard model recovering qualitatively the exact non-trivial solution. The formulation described below is based on Shankar's work (Shankar 1994).

11.1 Non-interacting Fermion Model Under RG: Scale Invariance and Fixed Points

Consider a 1-D model of non-interacting spinless fermions described by the hamiltonian:

$$H_0 = \sum_k \epsilon(k) \Psi^\dagger(k) \Psi(k) \tag{11.1}$$

J. Merino and A. L. Yeyati, *Many-Body Techniques in Condensed Matter Physics*, UNITEXT for Physics, https://doi.org/10.1007/978-3-031-55143-7_11

where $\epsilon(k) = -2t\cos(ka)$, representing electrons on a chain of lattice parameter a. We consider the system at half-filling. Since we are interested in the long wave lengths or low energies close to the Fermi energy we can linearize $\epsilon(k)$ around the $k_F = \pm\pi/2/$ Fermi points. Dividing the electrons in Right (R) with $k > 0$ and Left (L) with $k < 0$ movers we expand the cosines as:

$$- 2t\cos\left(\frac{\pi}{2}a + ka\right) \approx 2ta|k| = v_{FR}|k| = +v_F|k|$$

$$-2t\cos\left(-\frac{\pi}{2}a + ka\right) \approx -2ta|k| = v_{FL}|k| = -v_F|k| \tag{11.2}$$

and the non-interacting hamiltonian can be re-expressed as:

$$H_0 = \sum_{-\Lambda<k<\Lambda} v_{FR}k\Psi_R^\dagger(k)\Psi_R(k) + \sum_{-\Lambda<k<\Lambda} v_{FL}k\Psi_L^\dagger(k)\Psi_L(k) \tag{11.3}$$

where k is the momentum referred to the Fermi points $\pm k_F$.

We can express the above model through the partition function:

$$Z_0 = \int \mathcal{D}\bar\Psi\mathcal{D}\Psi e^{-S_0}, \tag{11.4}$$

where:

$$S_0 = \sum_\alpha \int_{-\infty}^{\infty} \frac{d\omega}{2\pi} \int_{-\Lambda}^{\Lambda} \frac{dk}{2\pi} \bar\Psi_\alpha(k,\omega)(i\omega - v_F\alpha k)\Psi_\alpha(k,\omega), \tag{11.5}$$

The above is the Gaussian action of the fermion model at low energies.

We now apply the RG approach to the present model illustrating the general procedure in a simple model. The main steps are:

(1) Integrate out the modes with momentum in the interval $-\Lambda/s \leq |k| \leq \Lambda$ as shown in the Fig. 11.1. Note that in the present case the lattice imposes a natural momentum cut-off, $\Lambda = \pi/2$ (since momentum is measured with respect to k_F).

The partition function reads:

$$Z = \int_{0<|k|<\Lambda/s} \mathcal{D}\bar\Psi\mathcal{D}\Psi \int_{\Lambda/s<|k|<\Lambda} \mathcal{D}\bar\Psi\mathcal{D}\Psi e^{-S_0[\bar\Psi,\Psi,\bar\psi,\psi]} \tag{11.6}$$

Since in this case the integration over the fast modes $\{\bar\psi,\psi\}$ is also a Gaussian integration, it just gives a numerical prefactor. Hence, we have:

$$Z_{eff} \propto Z \tag{11.7}$$

The numerical prefactor is irrelevant and can be neglected.

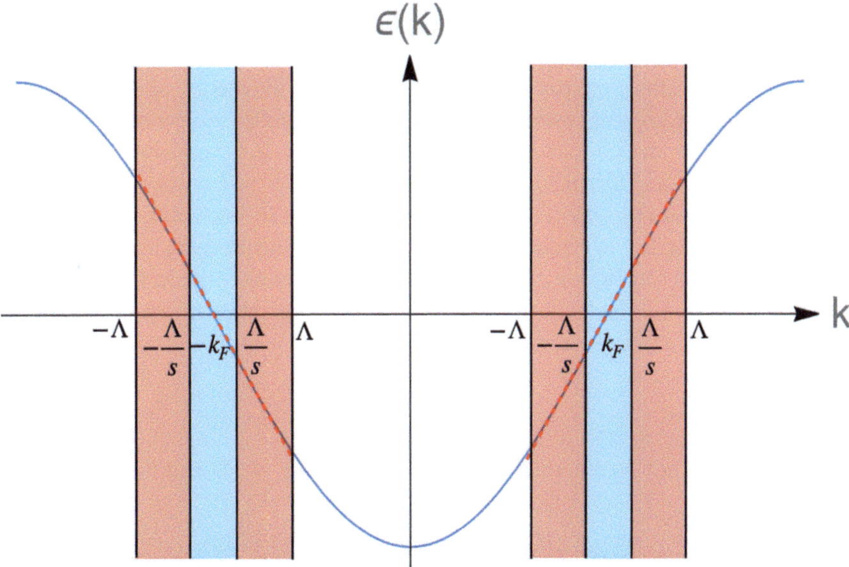

Fig. 11.1 Separation into fast (red regions) and slow (blue regions) modes, around the Fermi points, $\pm k_F$, in the 1D tight-binding model. The linear approximation to the tight-binding dispersion $\epsilon(k)$ around $\pm k_F$ used in the RG approach is shown (red dashed lines). The half-filled case, $\epsilon_F = 0$, is considered

(2) Rescaling. In order to compare a set of effective actions with the same cut-off (otherwise it would be meaningless to compare apples with oranges) the following change of variables or rescaling is performed:

$$k \to k' = sk$$
$$\omega \to \omega' = s\omega$$
$$\Psi(k, \omega) \to \Psi'(k', \omega') = s^{-3/2}\Psi(k, \omega) \qquad (11.8)$$

Introducing this change of variables in S_0 we find:

$$S_0 \to S_{eff} = \sum_\alpha \int \frac{d\omega'}{2\pi s} \int_{0<|k'|<\Lambda} \frac{dk'}{2\pi s} \bar{\Psi}'(k', \omega')s^{3/2}\frac{(i\omega' - v_{F\alpha}k')}{s}\Psi'(k', \omega')s^{3/2}$$
$$= S_0^*. \qquad (11.9)$$

where again the numerical prefactor associated with the integration over the fast modes has been ignored. Since the low momentum effective action $S_{eff} = S_0$ is invariant under the RG transformation the non-interacting fermion model described by S_0 is a fixed point. At a fixed point scale invariance is satisfied. This means that the correlation length, ξ, (quantifying the extension of the spatial fluctuations) either goes to zero, $\xi \to 0$, or diverges, $\xi \to \infty$. In this case the system has no preferred length scale displaying self-similarity as at a critical point in the theory of classical phase transitions. In the present system, $\xi \to \infty$, since the gap is zero for non-interacting fermions which is an example of an unstable fixed point. In the case in which $\xi \to 0$, the gap is infinity and the fixed point is stable.

We now consider the effect of a perturbation on the non-interacting fermion fixed point by applying the RG procedure. This will allow us to introduce the general procedure to treat interaction effects on the non-interacting fermion fixed point just discussed. This will allow illustrating the general procedure in a simple quadratic perturbation. We analyze the model:

$$H = H_0 + H_{pert} \tag{11.10}$$

where H_0 is the non-interacting fermionic hamiltonian in 1D and the perturbation reads:

$$H_{pert} = m \sum_j (-1)^j c_j^\dagger c_j \tag{11.11}$$

which describes a staggered mass term that doubles the unit cell when $m \neq 0$. Since H is quadratic it can be diagonalized exactly:

$$H = \sum_{k\alpha=\pm} \epsilon_\alpha(k) \Psi_\alpha^\dagger(k) \Psi_\alpha(k) \tag{11.12}$$

where the two dispersions are: $\epsilon_\pm(k) = \pm\sqrt{4t^2 cos^2(ka) + m^2}$. Note that for $m \neq 0$ the system opens a gap of magnitude $\sim |2m|$ at $\pm\pi/2$ leading to an insulating state at half-filling.

We are now analyzing the model from the RG approach point of view. Since we are interested on a low energy effective model, we only need to consider the excitations around the Fermi points. In order to do this we approximate the fermion operators as:

$$c_j = \frac{1}{\sqrt{N}} \sum_k e^{ikx_j} \Psi_k \approx e^{-ik_F x_j} \frac{1}{\sqrt{N}} \sum_{-k_F-\Lambda < k_L < -k_F+\Lambda} e^{ik_L x_j} \Psi(k_L)$$

$$+ e^{ik_F ja} \frac{1}{\sqrt{N}} \sum_{k_F-\Lambda < k_R < k_F+\Lambda} e^{ik_R x_j} \Psi(k_R) = e^{-ik_F x_j} \Psi_L(x_j) + e^{ik_F x_j} \Psi_R(x_j),$$

$$\tag{11.13}$$

with $k_F = \pi/2$ at half-filling and $x_j = ja$. While $e^{\pm ik_F ja}$ are rapidly oscillating in space, $\Psi_{R/L}(x_j)$ are slowly varying. The H_{pert} contribution can be re-expressed as:

$$H_{pert} = m \sum_j (-1)^j (\Psi_L^\dagger(x_j)\Psi_L(x_j) + \Psi_R^\dagger(x_j)\Psi_R(x_j)$$
$$+ e^{i2k_F ja}\Psi_L^\dagger(x_j)\Psi_R(x_j) + e^{-i2k_F ja}\Psi_R^\dagger(x_j)\Psi_L(x_j)).$$

$$(11.14)$$

Note that only for $k_F = \pi/2$ (half-filling) rapid oscillations disappear and the model reduces to:

$$H_{pert} = m \sum_j [(-1)^j (\Psi_L^\dagger(x_j)\Psi_L(x_j) + \Psi_R^\dagger(x_j)\Psi_R(x_j))$$
$$+ \Psi_L^\dagger(x_j)\Psi_R(x_j) + \Psi_R^\dagger(x_j)\Psi_L(x_j)].$$

$$(11.15)$$

The H_{pert} above is the starting point of the RG analysis. Let's consider the corresponding action for the cross terms mixing R/L and L/R movers:

$$S_{pert} = \int \frac{d\omega}{2\pi} \int \frac{dk}{2\pi} \{\bar{\Psi}_L(k,\omega)\Psi_R(k,\omega) + \bar{\Psi}_R(k,\omega)\Psi_L(k,\omega)\} \qquad (11.16)$$

The RG steps are:

(1) Integration over fast modes. Since the action is quadratic the integration over the fast modes would just lead to a constant that can be thrown away.
(2) Rescaling. We perform th evariable transformation:

$$k \to k' = sk$$
$$\omega \to \omega' = s\omega$$
$$\Psi(k,\omega) \to \Psi'(k',\omega') = s^{-3/2}\Psi(k,\omega) \qquad (11.17)$$

which leads to:

$$S_{pert} \to S_{pert}^{eff} = m \int \frac{d\omega}{2\pi s} \int \frac{dk}{2\pi s} s^3 \{\bar{\Psi}_L(k,\omega)\Psi_R(k,\omega) + \bar{\Psi}_R(k,\omega)\Psi_L(k,\omega)\}$$
$$= s S_{pert} \qquad (11.18)$$

Since the effective action of the slow modes increases under successive RG transformations, i.e. $S_{pert}^{eff} \to \infty$ as $s \to \infty$, this term is relevant in RG sense and will open a gap. This is indeed what one finds from the exact solution. Similarly, the diagonal L/L and R/R terms are also relevant under RG scaling.

Under the RG transformation the perturbation, S_{pert}, retains its form but with a multiplicative mass prefactor, $m(s)$, which depends on s as:

$$m(s) = sm(s = 1) \qquad (11.19)$$

The scaling parameter $s > 1$ is typically re-parameterised in the form: $s = e^t$ with $t \geq 0$. This means $t = ln(s)$ and $d\ln(s) = dt$. Introducing this change of variables in the above equation and deriving w.r.t t leads to the flow equation:

$$m(e^t) = e^t m(t = 0)$$

$$\frac{dm}{dln(s)} = sm(s = 1) = m(s).$$
$$(11.20)$$

This flow equation implies that if the mass parameter is initially positive $m(s \to 0^+) > 0^+, \frac{dm}{dlns} > 0$, and so $m(s)$ increases to infinitely large positive values while if initially $m(s \to 0^-) < 0^-$, $\frac{dm}{dlns} < 0$ and $m(s)$ decreases to infinitely large negative values. Hence, under the RG transformation the perturbation described by S_{pert} is relevant. The effective low energy model would then have a gap in agreement with the exact solution.

11.2   The One-Dimensional Hubbard Model Under RG: Spin-Charge Separation

We now apply the full RG machinery to the 1D Hubbard model whose solution is known exactly. We illustrate how the RG is in agreement with the exact solution of a highly non-trivial system.

The 1D HUbbard model reads:

$$H = H_0 + H_U = H_0 + U \sum_i n_{i\uparrow} n_{i\downarrow}.$$
$$(11.21)$$

The Hubbard contribution, H_U, in terms of fermion operators reads:

$$U n_{j\uparrow} n_{j\downarrow} = U c_{j\uparrow}^\dagger c_{j\uparrow} c_{j\downarrow}^\dagger c_{j\downarrow}.$$
$$(11.22)$$

It is convenient to re-express this in an equivalent but a spin symmetric form:

$$H_U = = U \sum_j (c_{j\alpha}^\dagger c_{j\alpha} - 1)^2 = U \sum_i (: c_{j\alpha}^\dagger c_{j\alpha} c_{j\beta}^\dagger c_{j\beta} : - : c_{j\alpha}^\dagger c_{j\alpha} : +1)$$

$$= U \sum_j (c_{j\alpha}^\dagger c_{j\beta}^\dagger c_{j\beta} c_{j\alpha} - c_{j\alpha}^\dagger c_{j\alpha} + 1)$$
$$(11.23)$$

where summing over the repeated α, β spin indices is assumed. The quadratic terms can be absorbed in the non-interacting ferrmion Gaussian action by readjusting the chemical potential and the constant term can be ignored. Hence, the Hubbard term effectively reduces to:

$$H_U \rightarrow U \sum_j c_{j\alpha}^\dagger c_{j\beta}^\dagger c_{j\beta} c_{j\alpha}. \tag{11.24}$$

which is our starting point for the RG analysis with the action:

$$S = S_0 + S_U = S_0 + \int dx d\tau \mathcal{L}_U(x, \tau), \tag{11.25}$$

with:

$$\mathcal{L}_U = U \bar{\Psi}_\alpha(x) \bar{\Psi}_\beta(x) \Psi_\beta(x) \Psi_\alpha(x). \tag{11.26}$$

Using the splitting of the fermion operator in fast and slow modes, the Hubbard term contains the following terms:

$$\mathcal{L}_U \rightarrow U[(\bar{\Psi}_{L\alpha}\Psi_{L\alpha})^2 + (\bar{\Psi}_{R\alpha}\Psi_{R\alpha})^2 + 2(\bar{\Psi}_{L\alpha}\Psi_{L\alpha})(\bar{\Psi}_{R\alpha}\Psi_{R\alpha})$$
$$+ 2(\bar{\Psi}_{L\alpha}\Psi_{R\alpha})(\bar{\Psi}_{R\beta}\Psi_{L\beta}) + e^{i4k_F x}(\bar{\Psi}_{L\alpha}\Psi_{R\alpha})^2 + e^{-i4k_F x}(\bar{\Psi}_{R\alpha}\Psi_{L\alpha})^2] \tag{11.27}$$

where the terms $\propto e^{i2k_F x}$ have been neglected since they will give negligible contribution to the integral in S_U due to the rapid spatial oscillations $e^{i2k_F x_j} = (-1)^j$ for $k_F = \pi/2a$. Note that at half-filling, we have $4k_F x_j = 2\pi$ so: $e^{i4k_F x_j} = e^{i2\pi j} = 1$. Such Umklapp process can occur since the momentum gained by the electronic system coincides with a reciprocal vector of the lattice $G = 4k_F = 2\pi/a$, so the total momentum of electrons and lattice is finally conserved in the Umklapp process.

In summary, the low energy processes due to the Coulomb interaction in 1D are decomposed in the scattering between either two R, two L electrons, a R and L electron, and two R(L) electrons scattered to the L(R) through Umklapp processes.

Since fermion fields occur as biquadratic terms it is convenient to rewrite the hamiltonian in terms of currents since all terms in the action are bilinear in the fermion fields. We define a charge current:

$$J_{L/R}(x) = \bar{\Psi}_{L/R\alpha}(x)\Psi_{L/R\alpha}(x) - \langle \bar{\Psi}_{L/R\alpha}(x)\Psi_{L/R\alpha}(x) \rangle_0, \tag{11.28}$$

where the average in the second term is over the non-interacting ground state. This sets the filled Fermi sea as the vacuum of excitations. The current operator is $U(1)$ symmetric since remains invariant under the transformation: $\Psi_{R/L} \rightarrow e^{i\theta}\Psi_{R/L}$. The average charge current in the ground state $\langle J_{L/R} \rangle_0 = 0$.

On the other hand, the spin current reads:

$$\mathbf{J}_{L/R}(x) = \frac{1}{2}\bar{\Psi}_{L/R\alpha}(x)\boldsymbol{\sigma}_{\alpha\beta}\Psi_{L/R\beta}(x), \tag{11.29}$$

the spin current is $SU(2)$ symmetric and satisfies $\langle \mathbf{J}_{L/R}(x) \rangle = 0$ in the ground state.

We can rewrite the hamiltonian in terms of charge and spin currents. For instance:

$$\mathbf{J}_L(x) \cdot \mathbf{J}_R(x) = \frac{1}{4}(\bar{\Psi}_{L\alpha}\boldsymbol{\sigma}_{\alpha\beta}\Psi_{L\beta}) \cdot (\bar{\Psi}_{R\gamma}\boldsymbol{\sigma}_{\gamma\delta}\Psi_{R\delta}) = \frac{1}{4}\boldsymbol{\sigma}_{\alpha\beta} \cdot \boldsymbol{\sigma}_{\gamma\delta}\bar{\Psi}_{L\alpha}\Psi_{L\beta}\bar{\Psi}_{R\gamma}\Psi_{R\delta}$$

$$= \frac{1}{4}(2\bar{\Psi}_{L\alpha}\Psi_{L\beta}\bar{\Psi}_{R\beta}\Psi_{R\beta} - \bar{\Psi}_{L\alpha}\Psi_{L\alpha}\bar{\Psi}_{R\delta}\Psi_{R\delta}) \tag{11.30}$$

where we have used the identity: $\boldsymbol{\sigma}_{\alpha\beta} \cdot \boldsymbol{\sigma}_{\gamma\delta} = 2\delta_{\alpha\delta}\delta_{\gamma\beta} - \delta_{\alpha\beta}\delta_{\gamma\delta}$. Hence we have:

$$\bar{\Psi}_L^\alpha \Psi_{R\alpha}\bar{\Psi}_R^\beta \Psi_{L\beta} = -2\mathbf{J}_L(x) \cdot \mathbf{J}_R(x) - \frac{1}{2}J_L(x)J_R(x). \tag{11.31}$$

The Hubbard interaction Lagrangian finally reads:

$$\begin{aligned}
\mathcal{L}_{int} &= U[J_L^2(x) + J_R^2(x) + J_L(x)J_R(x) - 4\mathbf{J}_L(x) \cdot \mathbf{J}_R(x) \\
&\quad + (\bar{\Psi}_L^\alpha \Psi_{R\alpha})^2 + (\bar{\Psi}_R^\beta \Psi_{L\beta})^2] \\
&= \frac{\pi}{2}\delta v_c(J_L^2(x) + J_R^2(x)) + \lambda_c J_L(x)J_R(x) + \lambda_s \mathbf{J}_L(x)\mathbf{J}_R(x) \\
&\quad + \lambda_U[(\bar{\Psi}_{L\alpha}\bar{\Psi}_{R\alpha})^2 + (\bar{\Psi}_{R\beta}\bar{\Psi}_{L\beta})^2],
\end{aligned}$$
$$\tag{11.32}$$

where we have defined the coupling constants describing the various terms as: $\frac{\pi}{2}\delta v_c = U, \lambda_c = U, \lambda_s = -4U, \lambda_U = U$.

11.2.1 RG Perturbative Treatment

In general, the RG cannot be applied exactly to a general interacting action but it can be implemented approximately based on perturbation theory. We will apply the RG to second order in the Coulomb interaction showing how already at this order it recovers the non-trivial physics of the 1D Hubbard model.

Our starting point is the action:

$$S = S_0^* + S_{int} \tag{11.33}$$

where the interacting part is typically a cuartic fermion interaction:

$$S \propto \int dx(\bar{\Psi}(x)\Psi(x))^2 = \int \frac{dk\,dq\,dp}{(2\pi)^6}\bar{\Psi}(p)\bar{\Psi}(q+k-p)\Psi(q)\Psi(k), \tag{11.34}$$

which can mix high and low energy modes.

The perturbative RG approach approximates the integration over the high energy modes in order to find the effective action of the slow modes through a straightforward perturbative expansion:

$$e^{-S_{eff}[\bar{\Psi},\Psi]} = \int \mathcal{D}\bar{\psi}\mathcal{D}\psi\, e^{-S[\bar{\Psi}+\bar{\psi},\Psi+\psi]}$$

$$= e^{-S_0^*[\bar{\Psi},\Psi]} \int \mathcal{D}\bar{\psi}\mathcal{D}\psi\, e^{-S_0^*[\bar{\psi},\psi]}(1 - S_{int}[\bar{\Psi}+\bar{\psi},\Psi+\psi]$$

$$+ \frac{1}{2!}S_{int}^2[\bar{\Psi}+\bar{\psi},\Psi+\psi] + \cdots).$$

$$(11.35)$$

Hence, when integrating out the high energy modes, the slow energy modes will be affected. Taking logarithms in the expression above the effective action can be obtained to a given order as:

$$S_{eff}[\bar{\Psi},\Psi] = S_0^* + \delta S^{(1)} + \delta S^{(2)} + \cdots$$

$$= S_0^*[\bar{\Psi},\Psi] - \ln\{\int \mathcal{D}\bar{\psi}\mathcal{D}\psi\, e^{-S_0^*[\bar{\psi},\psi]}(1 - S_{int}[\bar{\Psi}+\bar{\psi},\Psi+\psi]$$

$$+ \frac{1}{2!}S_{int}^2[\bar{\Psi}+\bar{\psi},\Psi+\psi] + \cdots)\}$$

$$\approx S_0^*[\bar{\Psi},\Psi] + \langle S_{int}[\bar{\Psi},\Psi]\rangle_0 - \frac{1}{2!}\langle S_{int}^2[\bar{\Psi},\Psi]\rangle_0 + \cdots$$

$$(11.36)$$

up to an irrelevant constant arising from adding and substracting the $\ln[\int \mathcal{D}\bar{\psi}\mathcal{D}\psi\, e^{-S_0^*[\bar{\psi},\psi]}$ term. Hence, the order n contribution to the effective action reads:

$$\delta S^{(n)} = (-1)^{n+1}\langle S_{int}^n[\bar{\Psi},\Psi]\rangle_0 = \frac{\int \mathcal{D}\bar{\psi}\mathcal{D}\psi\, e^{-S_0^*[\bar{\psi},\psi]}S_{int}^n[\bar{\Psi}+\bar{\psi},\Psi+\psi]}{\int \mathcal{D}\bar{\psi}\mathcal{D}\psi\, e^{-S_0^*[\bar{\psi},\psi]}}.$$

$$(11.37)$$

11.2.2 First Order Corrections

To first order in U we have to consider the terms associated to J_L^2, J_R^2, $J_L J_R$ and $\mathbf{J}_L \cdot \mathbf{J}_R$. These can be described diagrammatically as X-type diagrams with two incoming (Ψ fields) and two outcoming ($\bar{\Psi}$ fields) lines:

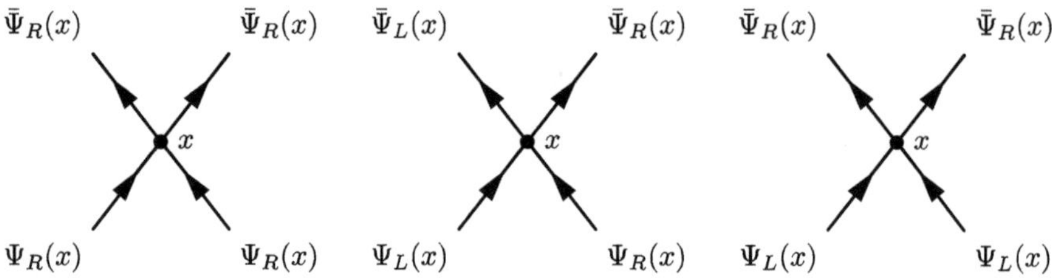

which correspond to $J_R(x)^2$ (left), $J_L(x)J_R(x)$ (middle), $\mathbf{J}_L \cdot \mathbf{J}_R(x)$ (middle) and Umklapp (right) processes.

Let's consider the behavior under RG of, for example, the contribution mixing R and L electrons to first order in U. Expressing the fields in terms of the fast and slow modes we have:

$$[(\bar{\Psi}_{R\alpha} + \bar{\Psi}_{R\alpha})(\Psi_{L\alpha} + \Psi_{L\alpha})]^2 = (\bar{\Psi}_{R\alpha}\Psi_{L\alpha})^2 + (\bar{\Psi}_{R\alpha}\Psi_{L\alpha})^2 + (\bar{\Psi}_{R\alpha}\Psi_{L\alpha})^2$$
$$+(\bar{\Psi}_{R\alpha}\Psi_{L\alpha})^2 + (\bar{\Psi}_{R\alpha}\Psi_{L\alpha})(\bar{\Psi}_{R\alpha}\Psi_{L\alpha}) + (\bar{\Psi}_{R\alpha}\Psi_{L\alpha})(\bar{\Psi}_{R\alpha}\Psi_{L\alpha}). \quad (11.38)$$

where we have neglected odd terms in $\bar{\Psi}, \Psi$ which give no contribution to $\delta S^{(1)} = \langle S_{int} \rangle_0$ due to symmetry in the integration. One can show that terms in (11.38) containing fast modes also give zero contribution except for terms involving two fast modes and two slow modes. For instance, the term:

$$\langle (\bar{\Psi}_{R\alpha}\Psi_{L\alpha})^2 \rangle_0 = 0. \quad (11.39)$$

The term $(\bar{\Psi}_{R\alpha}\Psi_{L\alpha})^2$ involving only slow modes, i.e., is just the original interaction unmodified by the integration over the fast modes. Performing the real space RG rescaling of these terms:

$$x' = \frac{x}{s}$$
$$\tau' = \frac{\tau}{s}$$
$$\bar{\Psi}'_{R/L}(x') = \bar{\Psi}_{R/L}(x)s^{1/2}$$
$$\Psi'_{R/L}(x') = \Psi_{R/L}(x)s^{1/2}$$

$$(11.40)$$

leads to:

$$\int dx d\tau (\bar{\Psi}_{R/L}\Psi_{R/L})^2 = \int dx' d\tau' (\bar{\Psi}'_{R/L}\Psi'_{R/L})^2. \quad (11.41)$$

Hence, the RG transformation leads to the equations $\lambda_c(s) = \lambda_c$, $\lambda_s(s) = \lambda_s$, i.e. these terms are marginal. The coupligns satisfy the equivalent set of RG flow equations:

$$\frac{d\lambda_c(s)}{d\ln(s)} = 0$$

$$\frac{d\lambda_s(s)}{d\ln(s)} = 0. \tag{11.42}$$

This means that, to first order in U, adding the λ_c and λ_s terms to S_0^* just shifts the non-interacting fixed point to another position in parameter space but the system remains qualitatively the same as for S_0^* i.e. it is gapless.

The integration over terms involving two fast modes and two slow modes leads to quadratic terms affecting S_0^*. The analysis of these terms are along the lines discussed previously to illustrate the full RG procedure. They can either lead to a chemical potential shift and/or a gap opening (staggered mass term).

In summary, the Hubbard interaction is marginal to first order in U. At this order we would conclude that the Hubbard model in 1D is essentially trivial containing the same qualitative physics as the non-interacting model. This is in contrast to the exact solution. We now analyze the behavior under RG of the Hubbard interaction at second order in U which leads to unusual physics beyond Fermi liquid theory.

11.2.3 Second Order Corrections

We now analyze higher order corrections to S_0^*. Specifically, we would like to know whether the second order term of the form:

$$\delta S^{(2)} = \frac{1}{2!}\langle S_{int}^2[\bar{\Psi} + \bar{\psi}, \Psi + \psi]\rangle_0 = \frac{1}{2!}\frac{\int \mathcal{D}\bar{\psi}\mathcal{D}\psi e^{-S_0^*[\bar{\psi},\psi]}S_{int}^2[\bar{\Psi} + \bar{\psi}, \Psi + \psi]}{\int \bar{\psi}\mathcal{D}\psi e^{-S_0^*[\psi]}},$$
$$\tag{11.43}$$

is relevant, irrelevant or marginal under the RG transformation.

The interaction term reads:

$$S_{int} = \int dx \Big(\frac{\pi}{2}\delta v_c(J_L(x)^2 + J_R(x)^2) + \lambda_c J_L(x)J_R(x) + \lambda_s \mathbf{J}_L(x) \cdot \mathbf{J}_R(x)$$
$$+ \lambda_U[(\bar{\Psi}_{L\alpha}\Psi_{R\alpha})^2 + (\bar{\Psi}_{R\alpha}\Psi_{L\alpha})^2]\Big) \tag{11.44}$$

where $x = (\mathbf{x}, \tau)$. The second order correction, $\delta S^{(2)}$, gives rise to many terms which can be analyzed separately.

Let's analyze first the cross $\lambda_c\lambda_U$ term:

$$\langle S_{int}^2[\bar{\Psi} + \bar{\slashed{\Psi}}, \Psi + \slashed{\Psi}]\rangle_0 = \lambda_c\lambda_U\langle\int dx\, J_L(x)J_R(x)\int dy((\bar{\Psi}_{L\gamma}(y)$$
$$+ \bar{\slashed{\Psi}}_{L\gamma}(y))(\Psi_{R\gamma}(y) + \slashed{\Psi}_{R\gamma}(y)))^2\rangle_0 \qquad (11.45)$$

Expressing it in terms of the fermion fields we can evaluate the $\langle ... \rangle_0$ using Wick's theorem on the fast fields, since we are averaging only over them in order to find the low energy effective theory. Let's consider a particular Wick contraction:

$$\langle S_{int}^2[\bar{\Psi} + \bar{\slashed{\Psi}}, \Psi + \slashed{\Psi}]\rangle_0$$
$$= \lambda_c\lambda_U\int dx\int dy\langle\bar{\Psi}_{L\alpha}(x)\slashed{\Psi}_{L\alpha}(x)\bar{\slashed{\Psi}}_{R\beta}(x)\Psi_{R\beta}(x)\bar{\slashed{\Psi}}_{L\gamma}(y)\slashed{\Psi}_{R\gamma}(y)\bar{\Psi}_{L\delta}(y)\Psi_{R\delta}(y)\rangle_0$$
$$= \lambda_c\lambda_U\int dx\int dy\,\bar{\Psi}_{L\alpha}(x)\Psi_{R\beta}(x)\bar{\Psi}_{L\delta}(y)\Psi_{R\delta}(y)\langle\slashed{\Psi}_{L\gamma}(y)\slashed{\Psi}_{L\alpha}(x)\rangle_0\langle\slashed{\Psi}_{R\beta}(x)\slashed{\Psi}_{R\gamma}(y)\rangle_0.$$
$$(11.46)$$

This particular choice corresponds to the following one-loop Feynman diagram in real space:

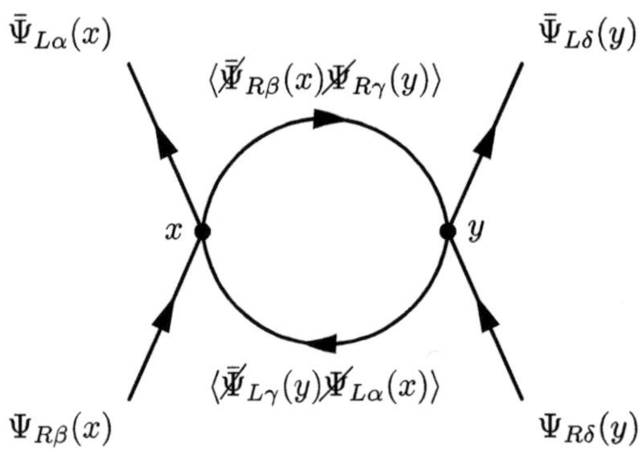

Again, note how Wick's theorem applies on the fast fermion fields. Since we are interested in the continuum limit, $a \to 0$, we can replace the slow fields at y by the ones at x. This is because we have:

$$\Psi_R(y) = \Psi_R(x) + (y - x)(\partial_\tau + i\partial_x)\Psi_R(x), \qquad (11.47)$$

but the term: $(y - x)(\partial_\tau + i\partial_x)\Psi_R(x) \to s\frac{1}{s}\frac{1}{\sqrt{s}}(\partial_\tau + i\partial_x)\Psi_R(x)$, so it is irrelevant under the RG transformation. So we can replace $\Psi_{R/L}(y) = \Psi_{R/L}(x)$ in the second order term which reads:

$$\langle S_{int}^2[\bar{\Psi}+\bar{\mathcal{W}},\Psi+\mathcal{W}]\rangle_0 = \lambda_c\lambda_U \int dx \int dy\,\bar{\Psi}_{L\alpha}(x)\Psi_{R\beta}\bar{\Psi}_{L\delta}(x)\Psi_{R\delta}(x)$$

$$\times \int \frac{dk}{(2\pi)^2}e^{ik(x-y)}\langle\bar{\mathcal{W}}_{L\gamma}(k)\mathcal{W}_{L\alpha}(k)\rangle_0 \int \frac{dk'}{(2\pi)^2}e^{ik'(y-x)}\langle\bar{\mathcal{W}}_{R\beta}(k')\mathcal{W}_{R\gamma}(k')\rangle_0.$$

$$= \lambda_c\lambda_U \int dx\,\bar{\Psi}_{L\alpha}(x)\Psi_{R\beta}\bar{\Psi}_{L\delta}(x)\Psi_{R\delta}(x)$$

$$\times \int \frac{dk}{(2\pi)^2}\int \frac{dk'}{(2\pi)^2} \int d(y-x)e^{i(k'-k)(y-x)}\frac{\delta_{\alpha\gamma}}{i\omega - v_F k}\frac{\delta_{\gamma\beta}}{i\omega' + v_F k'}$$

$$= \lambda_c\lambda_U \int dx\,\bar{\Psi}_{L\alpha}(x)\Psi_{R\beta}\bar{\Psi}_{L\delta}(x)\Psi_{R\delta}(x)\left[\frac{1}{(2\pi)^2}\int_{\Lambda/s<|k|<\Lambda}dk\int d\omega\frac{-\delta_{\alpha\beta}}{\omega^2 + v_F k^2}\right],$$

$$(11.48)$$

where the integration $\int$ is performed only over fast momenta $\Lambda/s < k < \Lambda$. We can now perform the integration explicitly:

$$-\int \frac{d\omega}{(2\pi)^2}\int_{\Lambda/s}^{\Lambda}\frac{dk}{\omega^2 + k^2} = -\int_{\Lambda/s}^{\Lambda}\frac{2\pi p\,dp}{(2\pi)p^2} = -\frac{1}{2\pi}lnp|_{\Lambda/s}^{\Lambda} = -\frac{1}{2\pi}\ln s.$$

$$(11.49)$$

where without loss of generality we assumed $v_F = 1$.

In summary, we have found that:

$$\langle S_{int}^2[\bar{\mathcal{W}}+\Psi]\rangle_0 \propto lns \int dx\,\bar{\Psi}_{L\alpha}(x)\Psi_{R\beta}(x)\bar{\Psi}_{L\delta}(x)\Psi_{R\delta}(x), \qquad (11.50)$$

which is an Umklapp term occurring in the original $S_{int}[\Psi]$ model action.

We now perform the rescaling in real space of the term:

$$x \to x' = xs$$
$$\tau \to \tau' = \tau s$$
$$\Psi \to \Psi' = s^{-1/2}\Psi \qquad (11.51)$$

and the $\delta S^{(2)}$ term behaves as:

$$s^2[\int \frac{dx}{s^2}\bar{\Psi}_{L\alpha}(x)\Psi_{R\beta}\bar{\Psi}_{L\delta}(x)\Psi_{R\delta}(x)]\ln s \qquad (11.52)$$

Thus, we find that the Umklapp coupling flows under the RG transformation as:

$$\lambda_U(s) \propto \lambda_c\lambda_U \ln s \qquad (11.53)$$

After including the four possible Wick contractions, the final RG flow equation for $\lambda_U(s)$ reads:

$$\frac{d\lambda_U}{d\ln s} = 4\lambda_c\lambda_U \qquad (11.54)$$

Applying the RG procedure to the other terms occurring in $\langle S_{int}^2 \rangle_0$ we arrive at the set of coupled RG equations:

$$\frac{d\lambda_c}{d\ln s} = 4\lambda_U^2$$

$$\frac{d\lambda_U}{d\ln s} = 4\lambda_c\lambda_U$$

$$\frac{d\delta v_c}{d\ln s} = 0$$

$$\frac{d\lambda_s}{d\ln s} = \lambda_s^2 \tag{11.55}$$

The first three equations describe the flow under RG of the couplings in the charge sector, the third the flows in the spin sector. The RG flow equation for δv_c shows that is marginal; when added to the fixed point the system remains metallic. So it will not be considered any further. The first important consequence of these equations is that the spin and charge sectors are decoupled since the RG equations (at least to order λ^2) do not mix the two. This prediction recovers the solution due to Bethe of the 1D Hubbard chain.

11.2.4   Spin Sector

The flow equation for the spin sector, $\frac{d\lambda_s}{d\ln s}$, implies that $\lambda_s(s)$ grows under the RG transformation. The solution is illustrated schematically in the RG flow diagram of Fig. 11.2. Integrating the flow equation we find:

$$\int \frac{d\lambda_s}{\lambda_s^2} = \int d\ln s \lambda(s) = \frac{-1}{\ln s + C}. \tag{11.56}$$

Where the constant C can be obtained from $\lambda_s(s = 1) = -4U \implies C = \frac{1}{4U}$. Depending on how we perturb the fixed point, S^*, two situations arise:

(i) If $U > 0$, the spin coupling is initially negative, $\lambda_s(s = 1) < 0$ and as s increases $\lambda(s) \to 0$. Hence, $\lambda(s)$ is irrelevant and the spin sector remains gapless as in the fixed point.

(ii) If $U < 0$, the spin coupling is initially positive, $\lambda_s(s = 1) > 0$ and as s increases $\lambda(s)$ increases. Hence, $\lambda(s)$ is relevant and the spin sector opens a gap.

λ_s

Fig. 11.2 RG flow diagram in the spin sector for the 1D Hubbard model

11.2.5 Charge Sector

The coupled set of flow equations read:

$$\frac{d\lambda_c}{d\ln s} = 4\lambda_U^2$$
$$\frac{d\lambda_U}{d\ln s} = 4\lambda_c\lambda_U$$

$$(11.57)$$

The RG flow diagram for the charge sector is shown in Fig.11.3. We can obtain the scaling equations from:

$$\frac{d}{d\ln s}(\lambda_c^2 - \lambda_U^2) = 2\lambda_c\frac{d\lambda_c}{d\ln s} - 2\lambda_U\frac{d\lambda_U}{d\ln s} = 0, \qquad (11.58)$$

which implies:

$$\lambda_c(s)^2 - \lambda_U(s)^2 = cte. \qquad (11.59)$$

The RG flow diagram of Fig. 11.3 is common to the Kondo model of magnetic impurities in metals and the Kosterlitz-Thouless transition in the 2D X-Y classical spin model.

11.2.6 Discussion of RG Analysis of the 1D Hubbard Model

We now combine the RG analysis in the charge and spin sectors to arrive at a full description of the physics of the 1D HUbbard model in 1D. We distinguish the two cases at half-filling:

(i) For $U > 0$: $\lambda_c, \lambda_U > 0$, $\lambda_s < 0$. Based on the RG flow diagram of Fig. 11.3, we have that $\lambda_c, \lambda_U \to \infty$. On the other hand $\lambda_s \to 0$. Thus, while a charge opens in the charge sector there is no spin-gap in the spin sector. This is an example of a Mott insulator with spin excitations consisting on gapless fermionic spinons.
(ii) For $U < 0$: $\lambda_c, \lambda_U < 0$, $\lambda_s < 0$. In this case $\lambda_c, \lambda_U \to 0$, $\lambda_s \to \infty$, so there is no charge gap but a spin gap. The system develops superconducting tendencies. Since the system is 1D Mermin-Wagner theorem forbids long range order even at $T = 0$.

Away from half-filling there is no Umklapp term in the model since $G \neq 4k_F$. Hence, in this case $\lambda_U = 0$. and the flow equation for λ_c reduces to:

$$\frac{d\lambda_c}{d\ln s} = 0. \qquad (11.60)$$

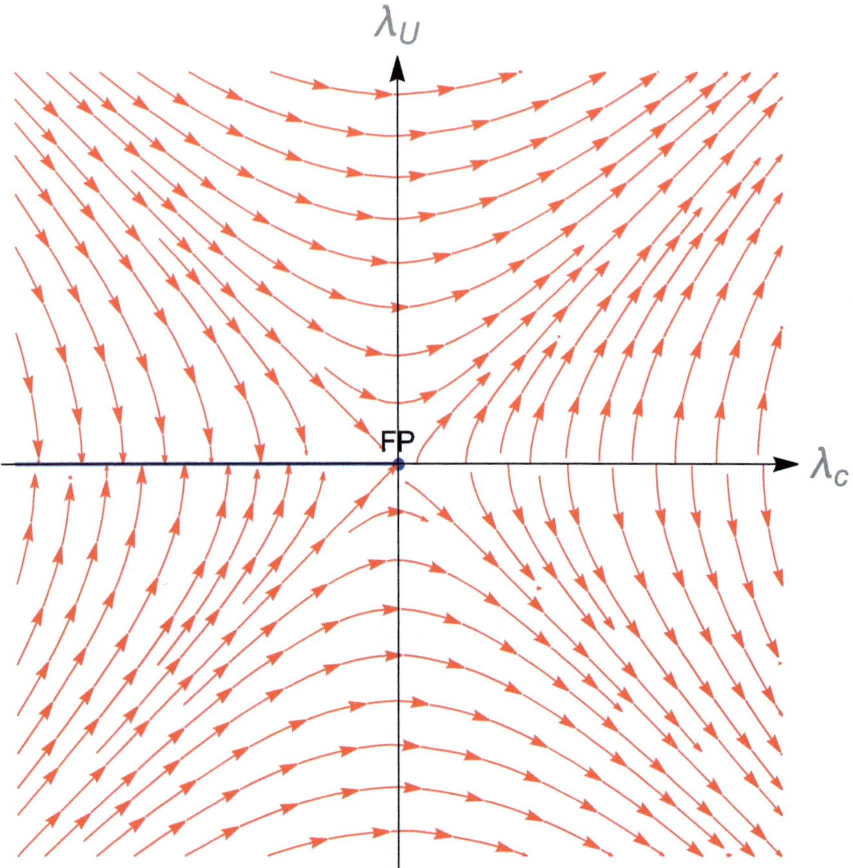

Fig. 11.3 RG flow diagram in the charge sector for the 1D Hubbard model. The blue line $\lambda_c < 0, \lambda_U = 0$ denotes a line of fixed points

which implies that in this case the Coulomb interaction is marginal and there is no gap opening. Hence, the system is metallic: a Luttinger liquid.

These results coincide with the exact Bethe ansatz solution (Lieb and Wu 1968) which, in particular, predicts that only at half-filling the 1D Hubbard model is insulating. Bosonization shows that spin-charge separation is present to all orders in λ.

Exercise III.13 RG approach to the 1D Hubbard model

A fundamental consequence of the RG analysis on the 1D Hubbard model is that spin and charge sectors are completely decoupled.

(i) Show that the cross charge-spin terms of the type: $\lambda_c \lambda_s$ do not arise explicitly in the RG flow equations of the 1D Hubbard model at $O(\lambda^2)$ due to a cancellation of the Wick contractions.

(ii) Obtain the prefactor of 4 in the Umklapp term λ_U in the RG flow Eq. (11.55).

Appendix
Hints for Solving Exercises

Part I: Equilibrium Many-Body Techniques

Exercise I.1: Hamiltonian of an interacting electron gas

(i)–(ii) First field operators are expressed in the plane wave basis:

$$\Psi_\sigma^\dagger(\mathbf{x}) = \frac{1}{\sqrt{V}} \sum_{\mathbf{k}} e^{-i\mathbf{k}\mathbf{x}} c_{\mathbf{k}\sigma}^\dagger$$

$$\Psi_\sigma(\mathbf{x}) = \frac{1}{\sqrt{V}} \sum_{\mathbf{k}} e^{i\mathbf{k}\mathbf{x}} c_{\mathbf{k}\sigma} \tag{A.1}$$

where the wavevectors $\mathbf{k} = \mathbf{p}/\hbar$. Introducing them in the hamiltonian one finds:

$$H = H_0 + V_{ee} = \sum_{\mathbf{k},\sigma} \frac{\hbar^2 k^2}{2m} c_{\mathbf{k},\sigma}^\dagger c_{\mathbf{k},\sigma} + \frac{1}{2V} \sum_{\alpha,\beta,\gamma,\delta} \sum_{\mathbf{k}_1,\mathbf{k}_2,\mathbf{k}_3,\mathbf{k}_4} V_{\mathbf{q}} c_{\mathbf{k}_4-\mathbf{q},\alpha}^\dagger c_{\mathbf{k}_3+\mathbf{q},\beta}^\dagger c_{\mathbf{k}_3+\mathbf{q}} c_{\mathbf{k}_3\gamma} c_{\mathbf{k}_4\gamma}, \tag{A.2}$$

with:

$$V_{\mathbf{q}} = \int d\mathbf{x}\, e^{i\mathbf{q}\cdot\mathbf{x}} \frac{e^2}{|\mathbf{x}|} = \frac{4\pi e^2}{q^2} \tag{A.3}$$

Since the Coulomb interaction does not change spin only the spin conserving terms are non-zero:

$$V_{ee} = \frac{1}{2V} \sum_{\sigma\sigma'} \sum_{\mathbf{k},\mathbf{k}',\mathbf{q}} V_{\mathbf{q}} c_{\mathbf{k}+\mathbf{q}\sigma}^\dagger c_{\mathbf{k}'-\mathbf{q}\sigma'}^\dagger c_{\mathbf{k}'\sigma'} c_{\mathbf{k}\sigma} \tag{A.4}$$

which explicitly shows the conservation of momentum and spin when two electrons are scattered. (iii) The normal ordering in the Coulomb interaction, V_{ee} means:

$$: n(\mathbf{x})n(\mathbf{x}') := \Psi^\dagger(\mathbf{x}')\Psi^\dagger(\mathbf{x})\Psi(\mathbf{x})\Psi(\mathbf{x}'). \tag{A.5}$$

© The Editor(s) (if applicable) and The Author(s), under exclusive license to Springer Nature Switzerland AG 2024

J. Merino and A. L. Yeyati, *Many-Body Techniques in Condensed Matter Physics*, UNITEXT for Physics, https://doi.org/10.1007/978-3-031-55143-7

Applying V_{ee} on the electron state $|\mathbf{x}\rangle$ to calculate the self-interaction contribution:

$$V_{ee}|\mathbf{x}\rangle = \frac{1}{2}\int d\mathbf{x}' \int d\mathbf{x}'' V(\mathbf{x}' - \mathbf{x}'')\Psi^\dagger(\mathbf{x}'')\Psi^\dagger(\mathbf{x}')\Psi(\mathbf{x}')\delta(\mathbf{x}'' - \mathbf{x}) = 0, \quad (A.6)$$

as it should. Had we not included the normal ordering in V_{ee} the self-interaction contribution would be non-zero. Hence, the Coulomb interaction can be expressed as:

$$V = \frac{1}{2}\int d\mathbf{x} \int d\mathbf{x}' V(\mathbf{x} - \mathbf{x}')\Psi^\dagger(\mathbf{x}')\Psi^\dagger(\mathbf{x})\Psi(\mathbf{x})\Psi(\mathbf{x}'). \quad (A.7)$$

Exercise I.2: Hubbard dimer

(i) The diagonalization of the hamiltonian, H, is greatly simplified using the $|S, S^z\rangle$ basis since H satisfies:

$$[S^z, H] = 0; \; [S^z, H] = 0 \quad (A.8)$$

where the total spin operators read:

$$S^\alpha = \frac{1}{2}\sum_i \sum_{\beta\gamma} c_{i\beta}^\dagger \sigma_{\beta\gamma}^\alpha c_{i\gamma}. \quad (A.9)$$

Since the hamiltonian does not mix sectors with different S, S^z quantum numbers, it is block diagonal in the $|S, S^z\rangle$ basis.

In addition the hamiltonian is symmetric under inversion w.r.t to the middle of the bond. Hence, parity, P, can also be used as a good quantum number to classify states:

$$[P, H] = 0. \quad (A.10)$$

$S = S^z = 0$ *sector.*
The states With even parity, $P = +1$ (even):

$$|\Psi_0\rangle = \frac{1}{\sqrt{2}}(c_{1\uparrow}^\dagger c_{2\downarrow}^\dagger - c_{1\downarrow}^\dagger c_{2\uparrow}^\dagger)|0\rangle$$

$$|\Psi_1\rangle = \frac{1}{\sqrt{2}}(c_{1\uparrow}^\dagger c_{1\downarrow}^\dagger + c_{2\uparrow}^\dagger c_{2\downarrow}^\dagger)|0\rangle \quad (A.11)$$

and with odd parity, $P = -1$ (odd) :

$$|\Psi_3\rangle = \frac{1}{\sqrt{2}}(c_{1\uparrow}^\dagger c_{1\downarrow}^\dagger - c_{2\uparrow}^\dagger c_{2\downarrow}^\dagger)|0\rangle \quad (A.12)$$

The hamiltonian for $P = +1$ reduces to:

$$H(S = S^z = 0, P = +1) = \begin{pmatrix} 0 & -2t \\ -2t & U \end{pmatrix} \quad (A.13)$$

The eeigenenergiees are:

$$E_{pm} = \frac{U}{2} \pm \frac{1}{2}\sqrt{U^2 + 16t^2} \tag{A.14}$$

The lowest ground state reads:

$$|\Psi_-\rangle = A(|\Psi_0\rangle + \beta|\Psi_1\rangle), \tag{A.15}$$

where:

$$A = \sqrt{\frac{1}{2}}\sqrt{1 + \frac{U/2}{\sqrt{(U/2)^2 + 4t^2}}}$$

$$\beta = \sqrt{\frac{1 - \frac{U/2}{\sqrt{(U/2)^2 + 4t^2}}}{1 + \frac{U/2}{\sqrt{(U/2)^2 + 4t^2}}}} \tag{A.16}$$

In the $U \to \infty$ limit:

$$\beta \to 0$$
$$A \to 1 \tag{A.17}$$

the ground state reads:

$$|\Psi_-(S = S^z = 0)\rangle = |\Psi_0\rangle = \frac{1}{\sqrt{2}}(c_{1\uparrow}^\dagger c_{2\downarrow}^\dagger - c_{1\downarrow}^\dagger c_{2\uparrow}^\dagger)|0\rangle \tag{A.18}$$

$S = 1,\ S^z = 0$ *sector.*
The only state is odd, $P = -1$: $|\Psi_4\rangle$ with energy, $E(S = 1,\ S^z = 0) = 0$ since $\langle\Psi_4|H|\Psi_4\rangle = 0$.
   $S = 1,\ S^z = \pm 1$ *sector.*
The two states:

$$|S = 1,\ S^z = +1\rangle = c_\uparrow^\dagger c_\uparrow^\dagger|0\rangle$$
$$|S = 1,\ S^z = -1\rangle = c_\downarrow^\dagger c_\downarrow^\dagger|0\rangle \tag{A.19}$$

have also $E(S = 1,\ S^z = \pm 1) = 0$.

Exercise I.3: Jordan-Wigner transformation

(i) We first reexpress the anisotropic XY model in terms of ladder operators:

$$H = \sum_i \frac{J_x + J_y}{4}\left(S_i^+ S_{i+1}^- + S_i^- S_{i+1}^+\right) + \frac{J_x - J_y}{4}\left(S_i^+ S_{i+1} + +S_i^- S_{i+1}^-\right) \tag{A.20}$$

We introduce the Jordan-Wigner transformation:

$$S_i^+ = d_i^+ e^{i\pi \sum_{l<i} d_l^\dagger d_l}$$

$$S_i^- = e^{-i\pi \sum_{l<i} d_l^\dagger d_l} d_i$$

$$S_i^z = d_i^\dagger d_i - \frac{1}{2} \tag{A.21}$$

so that the hamiltonian is transformed into spinless fermions. For instance, the term:

$$S_i^+ S_{i+1}^- = d_i^\dagger e^{-i\pi d_i^\dagger d_i} d_{i+1} = d_i^\dagger d_{i+1} \tag{A.22}$$

where in the last step we have used the fact that site i must be empty to have a non-zero value of the operator. Similarly:

$$S_i^- S_{i+1}^+ = d_i e^{i\pi d_i^\dagger d_i} d_{i+1}^\dagger = -d_i d_{i+1}^\dagger \tag{A.23}$$

On the other hand:

$$S_i^+ S_{i+1}^+ = d_i^\dagger e^{i2\pi \sum_{l<i} d_i^\dagger d_i} e^{i\pi d_i^\dagger d_i} d_{i+1}^\dagger = d_i^\dagger d_{i+1}^\dagger \tag{A.24}$$

and:

$$S_i^- S_{i+1}^- = d_{i+1} d_i . \tag{A.25}$$

So the hamiltonian can be expressed as:

$$H = -\sum t(d_i^\dagger d_{i+1} + d_{i+1}^\dagger d_i) + \Delta(d_{i+1}^\dagger d_i^\dagger + d_i d_{i+1}) \tag{A.26}$$

where: $t = \frac{J_x+J_y}{4}$, $\Delta = \frac{J_y-J_x}{4}$

(ii) It is convenient to express the hamiltonian in the momentum basis:

$$H = 2t \sum_{q>0} \epsilon_q (d_q^\dagger d_q - d_{-q} d_{-q}^\dagger) + \sum_{q>0} i \Delta_q (d_q^\dagger d_{-q}^\dagger + h.c) - \sum_{q>0} 2t\cos(qa) \tag{A.27}$$

where: $\epsilon_q = -2t\cos(q)$, $\Delta_q = 2\sin(qa)$. Omitting the constant term, the hamiltonian is expressed in matrix form as:

$$H = \sum_{q<0} \Psi_q^\dagger M(q) \Psi_q \tag{A.28}$$

where $\Psi_q^\dagger = (d_q^+, d_q)$ and:

$$M(q) = \begin{pmatrix} \epsilon_q & i\Delta_q \\ -i\Delta_q^* & -\epsilon_q \end{pmatrix} \tag{A.29}$$

In diagonal form:

$$H = \sum_Q \omega_q (a_q^\dagger a_q - a_{-q} a_{-q}^\dagger) \tag{A.30}$$

with the dispersion relation:

$$\omega_q = \sqrt{4t^2 \cos^2(qa) + 4\Delta^2 \sin^2(qa)} = \frac{1}{2}\sqrt{J_x^2 + J_y^2 + 2J_x J_y \cos(2qa)} \tag{A.31}$$

In the case $J_y = J_x$, spin excitations are gapless with the zero energy mode at $q = \pi/2$ and linear dispersion around this point. However, excitations are not magnons since, $\langle S^z \rangle = \langle n_f \rangle - 1/2 = 0$ since the system of spinless fermions is half-filled. Hence, the system has no long range order, it is a spin liquid whose elementary excitations are spinons.

In the case $J_y = 0$, the dispersion relation is independent of q, $\omega_q = J_x/2$, which corresponds to the 1D Ising ferromagnet.

Exercise I.4: Kitaev spin model

One should first notice that Majorana fermions satisfy the following anticommutation relation:

$$\{c_\alpha, c_\beta\} = 2\delta_{\alpha\beta} \tag{A.32}$$

since Majorana fermions are real: $c_\alpha^\dagger = c_\alpha$. i) Using the mapping of the Schwinger fermions onto Majorana fermions in the definition of the spin operators one finds:

$$S_j^x = \frac{i}{4}(b_j^x c_j - b_j^y b_j^z)$$
$$S_j^y = \frac{i}{4}(b_j^y c_j - b_j^z b_j^x)$$
$$S_j^z = \frac{i}{4}(b_j^z c_j - b_j^x b_j^y). \tag{A.33}$$

at any site j of the honeycomb lattice. Now we impose the constraints on the Schwinger fermions so that we can recover the original Hilbert space of the spins. Expressing the three constraints in terms of Majorana fermions we have:

$$b_j^x c_j + b_j^y b_j^z = 0$$
$$b_j^y c_j + b_j^z b_j^x = 0$$
$$b_j^z c_j + b_j^x b_j^y = 0 \tag{A.34}$$

which when introduced in thee expression for the spin operators (A.33) gives:

$$S_j^\alpha = \frac{i}{2}b^\alpha c. \tag{A.35}$$

Note that this expression incorporates the constraints exactly automatically recovering the original Hilbert space of the spins. Introducing the above expression in the original hamiltonian one finds:

$$H = i \sum_{\langle ij \rangle_\alpha} J_\alpha b_i^\alpha i b_j^\alpha c_i c_j = \sum_{\langle ij \rangle_\alpha} J_\alpha u_{ij}^\alpha c_i c_j. \tag{A.36}$$

where u_{ij}^α are bond fluxes connecting neighboring sites $\langle ij \rangle_\alpha$ in the α direction.

(ii) Since $[u_{ij}^\alpha, H] = 0$, the eigenstates of H are simultaneous eigenstates of u_{ij}^α. The eigenvalues of $_u ij^\alpha$ can only take two values, ± 1. It can be shown that the ground state of H is in the flux sector in which all $u_{ij}^\alpha = 1$. Hence, the ground state of the model is described by the quadratic hamiltonian:

$$H = i \sum_{\langle ij \rangle_\alpha} J_\alpha c_i c_j. \tag{A.37}$$

where $i \in A$ and $j \in B$ sites of the honeycomb lattice. This is a tight-binding model on a honeycomb lattice with positive pure imaginary hoppings going from A to B sites. When $J_x = J_y = J_z = K$ one recovers the band structure of graphene hosting Majorana fermions. The exact ground state of the Kitaev model is a gapless quantum spin liquid. Excitations consist of itinerant Majorana's in combination with flux vortices.

Exercise I.5: Green function of free phonons

The definition of the phonon propagator in momentum reads:

$$D(\mathbf{q}, t) \equiv -i \langle T [\phi(\mathbf{q}, t) \phi^+(q, 0)] \rangle \tag{A.38}$$

where the Fourier components of the displacement field read:

$$\phi(\mathbf{q}) = \sqrt{\frac{\hbar}{2m\omega_\mathbf{q}}} (b_\mathbf{q} + b_{-\mathbf{q}}). \tag{A.39}$$

Solving the Heisenberg EOM for the boson operators:

$$b_\mathbf{q}(t) = b_\mathbf{q} e^{-i\omega_\mathbf{q} t}$$
$$b_{-\mathbf{q}}^\dagger(t) = b_{-\mathbf{q}}^\dagger e^{i\omega_\mathbf{q} t}, \tag{A.40}$$

assuming $\omega_\mathbf{q} = \omega_{-\mathbf{q}}$. Introducing them in the now the displacement fields we have for the phonon propagator:

$$D(\mathbf{q}, t) \equiv -i \frac{\hbar}{2m\omega_\mathbf{q}} (\langle b_\mathbf{q} b_\mathbf{q}^\dagger \rangle e^{-i\omega_\mathbf{q} t} \theta(t) + \langle b_{-\mathbf{q}} b_{-\mathbf{q}}^\dagger \rangle e^{-i\omega_\mathbf{q} t} \theta(-t)) \tag{A.41}$$

And Fourier transforming:

$$D(\mathbf{q}, \nu) = \frac{\hbar}{2m\omega_{\mathbf{q}}} \left(\frac{1}{\nu - \omega_{\mathbf{q}} + i\delta} - \frac{1}{\nu + \omega_{\mathbf{q}} - i\delta} \right). \qquad (A.42)$$

Exercise I.6: Green function and ground state properties of a non-interacting electron gas

(i) Introducing the time-dependent fermion operators obtained from Heisenberg EOM and evaluating the anticommutator we have:

$$G_{\sigma}^{R}(\mathbf{k}, t - t') = -i\theta(t - t')e^{-i\epsilon_{\mathbf{k}}(t-t')} \qquad (A.43)$$

(ii) The Fourier transform is straightforward just introducing a convergence factor: $e^{-\delta t}$.

(iii) In the general interacting case, the Fourier transform of the retarded Green function can be performed introducing for $t > 0$ a semicircular contour for negative imaginary frequencies in the clockwise direction which encloses a single pole at $\epsilon_{\mathbf{k}} - \frac{i}{2\tau}$. Using the residue theorem one finds :

$$G_{\sigma}^{R}(\mathbf{k}, t) = -ie^{-i\epsilon_{\mathbf{k}}t}e^{-\frac{t}{2\tau}}. \qquad (A.44)$$

for $t > 0$, taking the time $t' = 0$ for simplicity. For $t < 0$ a semicircular contour is taken for positive imaginary frequencies which encloses no poles and so the integral gives 0. Thus, we have:

$$G_{\sigma}^{R}(\mathbf{k}, t) = -i\theta(t)e^{-i\epsilon_{\mathbf{k}}t}e^{-\frac{t}{2\tau}}. \qquad (A.45)$$

τ is the finite lifetime of the quasiparticles. The spectral density in this case is just a lorentzian of width $\frac{1}{2\tau}$ centered at $\epsilon_{\mathbf{k}}$. In the non-interacting case it is just a Dirac delta function describing a stationary state at energy $\epsilon_{\mathbf{k}}$.

(iv) By using the definitions one finds:

$$\langle n(\mathbf{x}) \rangle = (2S + 1)\frac{1}{V} \sum_{\mathbf{k}} \theta(k_F - k)$$

$$\langle T(\mathbf{x}) \rangle = (2S + 1)\frac{1}{V} \sum_{\mathbf{k}} \frac{\hbar^2 k^2}{2m} \theta(k_F - k), \qquad (A.46)$$

as it should.

Exercise I.7: Lehmann representation of the Hubbard dimer

The Lehmann representation of the Green function in real space reads:

$$G_{ii\sigma}(\omega) = \sum_{n} \left(\frac{|\langle \Psi_n(N+1)|c_{i\sigma}^{\dagger}|\Psi_G(N)\rangle|^2}{\omega - (E_n(N+1) - E_G(N)) + i\eta} + \frac{|\langle \Psi_n(N-1)|c_{i\sigma}|\Psi_G(N)\rangle|^2}{\omega + (E_n(N-1) - E_G(N)) - i\eta} \right).$$
$$(A.47)$$

At half-filling we need to first find the ground state, $E_G(2)$, of the Hubbard dimer with $N = 2$. We then need to obtain the whole energy spectra of the system with $N = 1$ and $N = 3$ electrons, $E_n(N = 1)$ and $E_n(N = 3)$. The resulting energies when adding an electron: $N \rightarrow N + 1$

n	$E_n(N = 3) - E_G(N = 1)$
0	$-t + U/2 + \sqrt{U^2 + 16t^2}/2$
1	$t + U/2 + \sqrt{U^2 + 16t^2}/2$

and when extracting an electron: $N \rightarrow N - 1$

n	$E_n(N = 2) - E_G(N = 1)$
0	$U/2 - \sqrt{U^2 + 16t^2}/2 + t$
1	$U/2 - \sqrt{U^2 + 16t^2}/2 - t$

The spectral function at a given site i can be obtained from:

$$\rho_{ii,\sigma}(\omega) = -\frac{1}{\pi} Im G_{ii\sigma}(\omega + i\eta), \tag{A.48}$$

which keads to a set of delta peaks.

In the limit $U \gg t$ the spectrum consists of four peaks: two peaks around $-U/2$ and two peaks around $U/2$ corresponding to the lower and upper Hubbard bands. Each pair of peaks is separated by $2t$, the effective "bandwidth" of the dimer.

For $U = 0$ the spectrum consists of two peaks centered at $\pm t$.

Exercise I.8: Interacting Fermi gas.

(i) You can have a look at Fig. 8, page 73 of the book of AGD for all the second order diagrams.

(ii) The Hartree diagram yields:

$$\Sigma^H(\mathbf{k}, \omega) = -2i \int \frac{d^4k_1}{(2\pi)^4} G^{(0)}(k_1) e^{i\omega_1 0^+} V(0) \tag{A.49}$$

while the Fock term is:

$$\Sigma^F(\mathbf{k}, \omega) = i \int \frac{d^4k_1}{(2\pi)^4} V(k - k_1) G^{(0)}(k_1) e^{i\omega_1 0^+} V(0) \tag{A.50}$$

Using:

$$i \int \frac{d\omega_1}{2\pi} e^{-i\omega_1 0^-} G^{(0)}(\mathbf{k}_1, \omega_1) = -\theta(k_F - k), \tag{A.51}$$

we finally have the first order correction for the self-energy:

$$\Sigma^{(1)}(\mathbf{k}, \omega) = V(0)n - \frac{e^2 k_F}{\pi} \left(1 + \frac{k_F^2 - k^2}{2kk_F} ln|\frac{k + k_F}{k - k_F}| \right). \tag{A.52}$$

The first term in the r.h.s can be absorbed in the chemical potential. The second term in the self-energy is real and independent of frequency and leads to a negative shift of the bare single electron energies:

$$\epsilon^*(\mathbf{k}) = \frac{\hbar^2 k^2}{2m} + \Sigma^{(1)}(\mathbf{k}) \tag{A.53}$$

Note that for a metal with density n corresponding to $\frac{r_s}{a_0} = 3.0146$, the energy is unshifted: $\epsilon^*(k_F) = \epsilon_F^0 = \frac{\hbar^2 k^2}{2m}$.

Exercise I.9: Quasiparticles in a Fermi liquid

(i) The expression of the second order diagram for the self-energy is:

$$\Sigma^{(2)}(\mathbf{k}, i\omega_n) =$$
$$-\frac{2}{\beta^2} \sum_{n_2, n_3} \int \frac{d\mathbf{k}_1}{(2\pi)^3} \int \frac{d\mathbf{k}_2}{(2\pi)^3} \int \frac{d\mathbf{k}_3}{(2\pi)^3} \frac{1}{i(\omega_n + \omega_{n_2} - \omega_{n_3}) - \epsilon_{\mathbf{k}_1}} \frac{1}{i\omega_{n_2} - \epsilon_{\mathbf{k}_2}} \frac{1}{i\omega_{n_3} - \epsilon_{\mathbf{k}_3}} \times$$
$$V(\mathbf{k}_2 - \mathbf{k}_3) V(\mathbf{k}_3 - \mathbf{k}_2) \delta(\mathbf{k}_1 - (\mathbf{k} + \mathbf{k}_2 - \mathbf{k}_3)) \tag{A.54}$$

We first perform the Matsubara sum over ω_{n_3}:

$$\frac{1}{\beta} \sum_{\omega_{n_3}} \frac{1}{i\omega_{n_3} - \epsilon_{\mathbf{k}_3}} \frac{1}{i(\omega_n + \omega_{n_2} - \omega_{n_3}) - \epsilon_{\mathbf{k}_1}} = (f(\epsilon_{\mathbf{k}_3}) - f(-\epsilon_{\mathbf{k}_1})) \frac{1}{i(\omega_n + \omega_{n_2}) - \epsilon_{\mathbf{k}_1} - \epsilon_{\mathbf{k}_3}} \tag{A.55}$$

Summing over ω_{n_2}:

$$\frac{1}{\beta} \sum_{\omega_{n_2}} \frac{1}{i\omega_{n_2} - \epsilon_{\mathbf{k}_2}} \frac{1}{i(\omega_n + \omega_{n_2}) - \epsilon_{\mathbf{k}_1} - \epsilon_{\mathbf{k}_3}} = (f(\epsilon_{\mathbf{k}_2}) + b(\epsilon_{\mathbf{k}_1} + \epsilon_{\mathbf{k}_3})) \frac{1}{i\omega_n + \epsilon_{\mathbf{k}_2} - \epsilon_{\mathbf{k}_1} - \epsilon_{\mathbf{k}_3}} \tag{A.56}$$

The imaginary part of the self-energy for $\omega \leq 0$ would then read:

$$\mathrm{Im}\,\Sigma^{(2)}(\mathbf{k}, \omega + i\eta) =$$
$$-2\pi \int \frac{d\mathbf{k}_1}{(2\pi)^3} \int \frac{d\mathbf{k}_2}{(2\pi)^3} \frac{d\mathbf{k}_3}{(2\pi)^3} \delta(\mathbf{k}_1 - (\mathbf{k} + \mathbf{k}_2 - \mathbf{k}_3)) \delta(\omega + \epsilon_{\mathbf{k}_2} - \epsilon_{\mathbf{k}_1} - \epsilon_{\mathbf{k}_3})$$
$$\theta(\epsilon_{\mathbf{k}_1}) \theta(\epsilon_{\mathbf{k}_3}) \theta(-\epsilon_{\mathbf{k}_2}) V(\mathbf{k}_3 - \mathbf{k}_2)^2.$$

(ii) Assuming a momentum independent (local) Coulomb interaction, $V(\mathbf{k}_3 - \mathbf{k}_2) = U$, we have:

$$\mathrm{Im}\,\Sigma^{(2)}(\mathbf{k}, \omega + i\eta) = -2\pi U^2 \int d\epsilon_2 \int d\epsilon_1 \int d\epsilon_3 \delta(\omega + \epsilon_2 - \epsilon_1 - \epsilon_3)\theta(\epsilon_1)\theta(\epsilon_3)\theta(-\epsilon_2)$$

$$\propto -\int_{-\omega}^{0} d\epsilon_2 \int_{0}^{\epsilon_2+\omega} d\epsilon_1 = -\int_{-\omega}^{0} d\epsilon_2(\epsilon_2 + \omega) \propto -\omega^2/2 \qquad (A.57)$$

Hence, we find the important result:

$$\mathrm{Im}\,\Sigma^{(2)}(\mathbf{k}, \omega + i\eta) \propto -\omega^2 \qquad (A.58)$$

as $k \to k_F$. Luttinger showed that this ω^2 dependence survives to any order in perturbation theory in the $\omega \to 0$ limit.

(iii) This exercise is solved in the lectures.

iv) Using the Matsubara representation of the imaginary Green function:

$$G_\sigma(\mathbf{k}, 0^-) = \frac{1}{\beta}e^{i\omega_n 0^+}G_\sigma(\mathbf{k}, i\omega_n) = \frac{1}{\beta}\sum_{\omega_n} e^{i\omega_n 0^+}\frac{Z_\mathbf{k}}{i\omega_n - \epsilon_\mathbf{k}^*} = Z_\mathbf{k} f(\epsilon_\mathbf{k}^*), \quad (A.59)$$

valid for $k \to k_F$. In the limit $T \to 0$ this expression leads to a discontinuity of size $Z_\mathbf{k}$ at the Fermi energy, $\epsilon_F^* = \epsilon_{\mathbf{k}_F}^*$ in the occupation number.

Exercise I.10: Finite temperature Green function for phonons

Solving the Heisenberg EOM for the boson operators in imaginary times, we have:

$$b^\dagger(\tau) = e^{\omega\tau}b(0)$$
$$b(\tau) = e^{-\omega\tau}b(0). \qquad (A.60)$$

Introducing these in $x(\tau)$, the phonon propagator reads:

$$D(\tau) = -\frac{\hbar}{2m\omega}(e^{-\omega\tau}(1 + b(\omega)) + e^{\omega\tau}b(\omega). \qquad (A.61)$$

Performing the Fourier transform to Matsubara frequencies we find:

$$D(i\nu_n) = \int_{0}^{\beta} e^{i\nu_n\tau}D(\tau)d\tau = -\frac{\hbar}{2m\omega}\frac{2\omega}{\omega^2 + \nu_n^2}. \qquad (A.62)$$

Analytical continuation to real frequencies, $i\nu_n \to \nu + i\eta$ gives:

$$D(\nu) = \frac{\hbar}{m}\frac{1}{\nu^2 - \omega^2}. \qquad (A.63)$$

Exercise I.11: Electron-phonon interaction in solids

(i) The effective interaction between two electrons mediated by a phonon is given by:

$$V_{eff}(\mathbf{q}, i\nu_n) = g_{\mathbf{q}}^2 \frac{2\omega_{\mathbf{q}}}{\nu^2 - \omega_{\mathbf{q}}^2} \qquad (A.64)$$

Since the characteristic phonon frequency scale in a solid is given by the Debye frequency, $\omega_{\mathbf{q}} \sim \omega_D$. So at sufficiently low frequencies, $\nu < \omega_D$ the effective interaction between electrons is attractive allowing for superconducting instabilities!

(ii) The leading order Feynman diagram for the effect of the the electron-phonon interaction on electron propagation is:

$$\Sigma(k) = \frac{1}{V} \sum_{\mathbf{q}} (ig_{\mathbf{q}})^2 G^{(0)}(k - q) D(q) \qquad (A.65)$$

which explicitly in terms of Matsubara frequencies reads

$$\Sigma(\mathbf{k}, i\omega_n) = -\frac{1}{V\beta} \sum_{\mathbf{q}, i\omega_n} g_{\mathbf{q}}^2 \frac{2\omega_{\mathbf{q}}}{(i\nu_n)^2 - \omega_{\mathbf{q}}^2} \frac{1}{i\omega_n - i\nu_n - \epsilon_{\mathbf{k}-\mathbf{q}}}. \qquad (A.66)$$

(iii) The Matsubara sums can be performed by first splitting the expression for the self-energy obtained in b) in two contributions:

$$\Sigma(\mathbf{k}, i\omega_n) = -\frac{1}{V\beta} \sum_{\mathbf{q}, i\omega_n} g_{\mathbf{q}}^2 \left(\frac{1}{i\nu_n - \omega_{\mathbf{q}}} \frac{1}{i\omega_n - i\nu_n - \epsilon_{\mathbf{k}-\mathbf{q}}} - \frac{1}{i\nu_n - \omega_{\mathbf{q}}} \frac{1}{i\omega_n - i\nu_n - \epsilon_{\mathbf{k}-\mathbf{q}}} \right). \qquad (A.67)$$

Performing the Matsubara sums over the two contributions one finds:

$$\Sigma(\mathbf{k}, i\omega_n) = \frac{1}{V} \sum_{\mathbf{q}} g_{\mathbf{q}}^2 \left(\frac{1 + b(\omega_{\mathbf{q}}) - f(\epsilon_{\mathbf{k}-\mathbf{q}})}{i\omega_n - (\omega_{\mathbf{q}} + \epsilon_{\mathbf{k}-\mathbf{q}})} + \frac{b(\omega_{\mathbf{q}}) + f(\epsilon_{\mathbf{k}-\mathbf{q}})}{i\omega_n - (\epsilon_{\mathbf{k}-\mathbf{q}} - \omega_{\mathbf{q}})} \right) \qquad (A.68)$$

While the first term describes the emission of a phonon by an electron propagating along the solid, the second term describes the absorption of a phonon by a hole.

Exercise I.12: Charge order instabilities

(i) The RPA charge susceptibility reads:

$$\chi_c(q) = \frac{\chi^{(0)}(q)}{1 + V(\mathbf{q})\chi^{(0)}(q)}, \qquad (A.69)$$

where the Coulomb repulsion in the extended Hubbard model on the square lattice is: $V(\mathbf{q}) = U/2 + V(\cos(q_x) + \cos(q_y))$. The bare susceptibility is:

$$\chi^{(0)}(\mathbf{q}, i\nu) = -\frac{2}{V} \sum_{\mathbf{k}} \frac{f(\epsilon_{\mathbf{k}}) - f(\epsilon_{\mathbf{k}+\mathbf{q}})}{i\nu + \epsilon_{\mathbf{k}} - \epsilon_{\mathbf{k}+\mathbf{q}}} \qquad (A.70)$$

with $\epsilon_{\mathbf{k}} = -2t(\cos(k_x) + \cos(k_y))$, the electron tight binding dispersion. The condition for a charge order instability with modulation $\mathbf{Q}$ reads:

$$V(\mathbf{Q})\chi^{(0)}(\mathbf{Q}, 0) = -1, \tag{A.71}$$

which is possible since $V(\mathbf{q})$ can be negative although $U, V > 0$. For instance, at half-filling, the Fermi surface is nested since $\mathbf{Q} = (\pi/a, \pi/a, \pi/a)$ coincides with $\frac{\mathbf{G}}{2}$ where $\mathbf{G}$ is a reciprocal vector of the lattice. Hence, since, at half-filling: $\epsilon_{\mathbf{p}+\mathbf{Q}} = -\epsilon_{\mathbf{p}}$, $\chi^{(0)}(\mathbf{Q}, \omega = 0) \to \infty$. An instability with modulation $\mathbf{Q}$ occurs when

$$V(\mathbf{Q}) = -1/\chi^{(0)}(\mathbf{Q}, 0) \to 0^-. \tag{A.72}$$

which implies, $V(\mathbf{Q}) = U/2 - 3V = 0^-$. Thus, for $V \gtrsim U/6$ a charge order instability corresponding to charge alternation occurs in the extended Hubbard model on the square lattice.

(ii) The evaluation of the self-energy diagram is similar to that of the electron self-energy in the electron-phonon problem. After performing Matsubara sums one finds:

$$\Sigma(\mathbf{k}, i\omega_n) = \frac{1}{V}\sum_{\mathbf{q}} V(\mathbf{q})^2 \int_0^\infty Im\chi_c(\mathbf{q}, \nu) \left(\frac{1 + b(\nu) - f(\epsilon_{\mathbf{k}-\mathbf{q}})}{i\omega_n - (\nu + \epsilon_{\mathbf{k}-\mathbf{q}})} + \frac{b(\nu) + f(\epsilon_{\mathbf{k}-\mathbf{q}})}{i\omega_n - (\epsilon_{\mathbf{k}-\mathbf{q}} - \nu)} \right). \tag{A.73}$$

This self-energy describes the effect of charge fluctuations on the propagation of an electron in the lattice.

Exercise I.13: Magnetic instabilities of the Hubbard model

(i) The critical U_c for ferromagnetism can be obtained from the RPA expression for the magnetic suscpetibility:

$$\chi^{+-}(q) = \frac{\chi^{(0)+-}(q)}{1 - U\chi^{(0)+-}(q)}. \tag{A.74}$$

which leads to the instability criteria:

$$U_c = \frac{1}{\chi^{(0)}(\mathbf{q}, 0)}, \tag{A.75}$$

where $\chi^{(0)}(\mathbf{q}, 0)$ is just the bare propagator of a particle-hole pair of the paramagnetic system. For ferromagnetism at $T = 0$, $\chi^{(0)}(\mathbf{q} \to \mathbf{0}, 0) = N_\sigma(\epsilon_F)$ and we recover the Stoner criteria:

$$U_c = \frac{1}{N_\sigma(\epsilon_F)}. \tag{A.76}$$

(ii) The Néel antiferromagnetic instability occurs at the wavevector: $\mathbf{Q} = (\pi/a, \pi/a, \pi/a)$ which coincides with $\frac{\mathbf{G}}{2}$. Thus, at half-filling: $\epsilon_{\mathbf{p}+\mathbf{Q}} = -\epsilon_{\mathbf{p}}$ for all $\mathbf{p}$ on the

Fermi surface. Hence, we expect that, at half-filling, any small perturbation will turn the paramagnetic state in the Hubbard model on the cubic lattice into a Néel state. This is readily seen by evaluating:

$$\chi^{(0)}(\mathbf{Q}, \omega = 0) = -\int_{-6t}^{\epsilon_F} \frac{N(\epsilon)}{2\epsilon} d\epsilon, \tag{A.77}$$

which diverges at half-filling, $\epsilon_F = 0$. This implies:

$$U_c = \frac{1}{\chi^{(0)}(\mathbf{Q}, 0)} \to 0^+ \tag{A.78}$$

as expected from the nesting of the Fermi surface with vector.

Part II: Non-equilibrium Many-Body Techniques

Exercise II.1: Single level GFs

The exercise is straightforward using $a(t) = ae^{-i\epsilon t}$, $a^\dagger(t) = a^\dagger e^{i\epsilon t}$ and $\langle a^\dagger a \rangle = n_F(\epsilon) = 1/(e^{\beta(\epsilon-\mu)} + 1)$ for fermions and $\langle a^\dagger a \rangle = n_B(\epsilon) = 1/(e^{\beta(\epsilon-\mu)} - 1)$ for bosons. In part vi, the relation

$$G^{+-}(\omega) + G^{-+}(\omega) = (2n_F - 1)\left(G^A(\omega) - G^R(\omega)\right)$$

for fermions and

$$G^{+-}(\omega) + G^{-+}(\omega) = (2n_B + 1)\left(G^A(\omega) - G^R(\omega)\right)$$

for bosons are only valid in equilibrium conditions.

Exercise II.2: Diagrams in Keldysh formalism

(i) At first order only the Hartree and Fock diagrams (first two diagrams in the upper panel of Fig. 8.2) contribute. Using the rules for constructing diagrams (see Sect. 3.7) for the Hartree diagram we find

$$G_{\text{Hartree}}^{(1),\alpha\beta}(x, x') =$$

$$-i \sum_\gamma \int dx_1 \int dx_2 g^{\alpha\gamma}(x, x_1) g^{\gamma\gamma}(x_2, x_2) g^{\gamma\beta}(x_1, x') v(r_1 - r2)\delta(t_1 - t - 2)(-1)^\gamma,$$

$$\tag{A.79}$$

while for the Fock contribution

$$G_{\text{Fock}}^{(1),\alpha\beta}(x, x') =$$

$$i\sum_{\gamma}\int dx_1 \int dx_2 g^{\alpha\gamma}(x, x_1)g^{\gamma\gamma}(x_1, x_2)g^{\gamma\beta}(x_2, x')v(r_1 - r2)\delta(t_1 - t - 2)(-1)^{\gamma} ,$$

$$(A.80)$$

(ii) As an example of disconnected diagram let us consider the following one

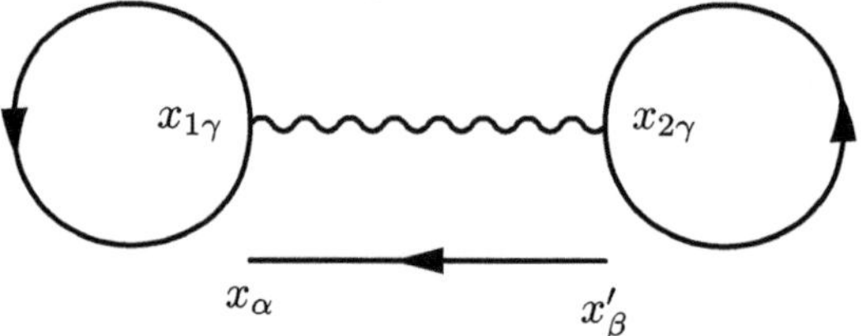

whose expression is

$$i\sum_{\gamma}\int dx_1 \int dx_2 g^{\gamma\gamma}(x_1, x_1)g^{\gamma\gamma}(x_2, x_2)g^{\alpha\beta}(x, x')v(r_1 - r_2)\delta(t_1 - t_2)(-1)^{\gamma} .$$

Taking into account that $g^{++}(x, x) = g^{--}(x, x)$ the terms in $\gamma = +$ and $\gamma = -$ cancel.

Exercise II.3: Triangular representation

(i) One can first assume that the rotated self-energy $\hat{U}\hat{\Sigma}\hat{U}^\dagger$ has a general form

$$\hat{U}\hat{\Sigma}\hat{U}^\dagger = \begin{pmatrix} \Sigma_1 & \Sigma_2 \\ \Sigma_3 & \Sigma_4 \end{pmatrix}$$

Then, the rotated second term in the Dyson equation $\hat{U}\hat{g} \otimes \hat{\Sigma} \otimes \hat{G}\hat{U}^\dagger$ would be given by

$$\hat{U}\hat{g} \otimes \hat{\Sigma} \otimes \hat{G}\hat{U}^\dagger = \begin{pmatrix} g^A\Sigma_4 G^R & g^A\Sigma_3 G^A + g^A\Sigma_4 G^K \\ g^R\Sigma_2 G^R & g^R\Sigma_1 G^A + g^R\Sigma_2 G^K + g^K\Sigma_3 G^A + g^K\Sigma_4 G^K \end{pmatrix}$$

where internal convolution products are implicitly assumed on the rhs. Conservation of the triangular structure thus requires $\Sigma_4 = (\Sigma^{++} + \Sigma^{+-} + \Sigma^{-+} + \Sigma^{--})/2 = 0$ or, equivalently $\Sigma^{++} + \Sigma^{--} = -(\Sigma^{-+} + \Sigma^{-+})$.

Exercise II.4: Langreth rules

Let's take the case of the retarded components. One has $(A \otimes B)^R \equiv (A \otimes B)^{-+} - (A \otimes B)^{--}$ which, after expanding and regrouping terms can be written as

$$(A \otimes B)^R = \int_{-\infty}^{+\infty} dt_1 A^{-+}(t, t_1)\left(B^{++}(t_1, t') - B^{+-}(t_1, t')\right) - A^{--}(t, t_1)\left(B^{-+}(t_1, t') - B^{--}(t_1, t')\right) .$$

Taking into account that $B^R = B^{++} - B^{+-} = B^{-+} - B^{--}$ one easily demonstrate the expected result.

Exercise II.5: Dyson equation for G^{+-}

Starting from the full Dyson equation $G = g(I + \Sigma G)$, where integration over internal time arguments along the Keldysh contour is implicitly assumed, using the Langreth rules one obtains

$$
\begin{aligned}
G^{+-} &= g^R(I + \sigma G)^{+-} + g^{+-}(I + \Sigma G)^A \\
&= g^R(\Sigma^R G^{+-} + \Sigma^{+-} G^A) + g^{+-}(I + \Sigma^A G^A) .
\end{aligned}
\tag{A.81}
$$

Operating along the lines in Sect. 8.5 for obtaining the Dyson equation for G^K (c.f. Eq. (8.35)) one reaches the expected result.

Exercise II.6: Uncoupled leads GFs

Substitution of Eq. (9.14) into Eq. (9.15) yields directly expression (9.17) for the boundary GF in a semi-infinite chain. Alternative, from the equation of motion for the semi-infinite chain GFS

$$
\begin{aligned}
g_{00} &= g_0 + g_0 V g_{10} \\
g_{10} &= g_{11} V g_1 ,
\end{aligned}
$$

where $g_0 = g_1 = 1/\omega$ are the uncoupled GFs for the isolated sites. Then, from symmetry considerations $g_{00} = g_{11} = g$ one readily obtains

$$
g = \left[\omega - V^2 g\right]^{-1}
$$

from which Eq. (9.17) follows.

Exercise II.7: Single channel contact transmission

Clearly $\tau(\omega)$ in Eq. (9.23) satisfies $\tau(\omega) \geq 0$ for $V' \in \mathbb{R}$ (if $V' \in \mathbb{C}$, V' in the numerator of Eq. (9.23) should be replaced by $|V'|$). Let's define $g^A_{L,R}(\omega) = z_{1,2} = a_{1,2} + ib_{1,2}$, with $b_{1,2} > 0$. Equation (9.23) can then be written as

$$
\tau = \frac{4V'^2 b_1 b_2}{1 - 2V'^2 (a_1 a_2 - b_1 b_2) + V'^4 |z_1|^2 |z_2|^2} .
$$

Moreover, one can check that $1/\tau$ reaches a minimum for $V_0' = \pm 1/\sqrt{|z_1 z_2|}$, given by

$$
1/\tau \rfloor_{V'=V_0'} = \frac{1 + |z_1 z_2| + b_1 b_2 - a_1 a_2}{2 b_1 b_2} \geq 1 ,
$$

which demonstrates that $\tau(\omega) \leq 1$ for any set of parameters.

Exercise II.8: Current noise in single channel contact

Analysis of different limits

- $T \to 0$ limit. In this limit the Fermi distribution functions become step Heaviside functions, i.e.: $n_{L,R}(\omega) = \theta(\mu_{L,R} - \omega)$ and $1 - n_{L,R}(\omega) = \theta(\omega - \mu_{L,R})$ Then

$$
\begin{aligned}
S(0) &= \frac{4e^2}{h} \int d\omega' \left\{ \tau \left[n_L(1 - n_R) + n_R(1 - n_L) \right] - \tau^2(n_R - n_L)^2 \right\} \\
&= \frac{4e^2}{h} \int_{\min \mu_L, \mu_R}^{\max \mu_L, \mu_R} d\omega' \left(\tau - \tau^2 \right) = \frac{4e^2}{h} \tau(1 - \tau)|\mu_L - \mu_R| = \frac{4e^2}{h} \tau(1 - \tau)eV
\end{aligned}
\tag{A.82}
$$

Thus, the shot noise result is recovered, $S(0) \propto \tau(1 - \tau)eV$.

- $V = 0$ limit. In this limit, $n_L(\omega) = n_R(\omega) \equiv n(\omega)$.

$$
S(0) = \frac{4e^2}{h} \int d\omega' 2\tau n(1 - n)
\tag{A.83}
$$

Then, using the property

$$
\frac{d}{d\omega} n = -\beta n(1 - n)
\tag{A.84}
$$

one finds

$$
\begin{aligned}
S(0) &= \frac{4e^2}{h} \int d\omega' 2\tau n(1 - n) = -\frac{8e^2\tau}{\beta h} \int_{-\infty}^{\infty} d\omega' \frac{d}{d\omega'} n(\omega') = -\frac{8e^2\tau}{\beta h} \left[n(\infty) - n(-\infty) \right] \\
&= -\frac{8e^2\tau}{\beta h}(0 - 1) = \frac{8e^2\tau}{h} k_B T
\end{aligned}
\tag{A.85}
$$

Thus, in this limit one obtains the thermal noise expression $S(0) \sim 4G_N k_B T$, where G_N is the normal conductance.

Exercise II.9: Full counting statistics

(i) Our aim is to evaluate the expression

$$
S(\chi, t_0) = \frac{t_0}{2\pi} \int d\omega \log \frac{\det \left(\mathbb{I} - \hat{V}'(\chi)\hat{g}_R \hat{V}'^{\dagger}(\chi)\hat{g}_L \right)}{\det \left(\mathbb{I} - \hat{V}'(0)\hat{g}_R \hat{V}'^{\dagger}(0)\hat{g}_L \right)}
$$

in the low energy approximation. We start by evaluating the matrix $\mathbb{I} - \hat{V}'\hat{g}_R\hat{V}'^{\dagger}\hat{g}_L$. In the full Keldysh representation we have

$$\hat{\mathcal{V}}' = \mathcal{V}'\begin{pmatrix} e^{i\chi/2} & 0 \\ 0 & -e^{-i\chi/2} \end{pmatrix} \qquad\qquad \hat{g}_{L,R} = \begin{pmatrix} g_{L,R}^{++} & g_{L,R}^{+-} \\ g_{L,R}^{-+} & g_{L,R}^{--} \end{pmatrix}$$

Then,

$$\mathbb{I} - \hat{\mathcal{V}}'\hat{g}_R\hat{\mathcal{V}}'^{\dagger}\hat{g}_L = \begin{pmatrix} 1 - \mathcal{V}'^2(g_R^{++}g_L^{++} - e^{i\chi}g_R^{+-}g_L^{-+}) & -\mathcal{V}'^2(g_R^{++}g_L^{+-} - e^{i\chi}g_R^{+-}g_L^{--}) \\ -\mathcal{V}'^2(g_R^{--}g_L^{-+} - e^{-i\chi}g_R^{-+}g_L^{++}) & 1 - \mathcal{V}'^2(g_R^{--}g_L^{--} - e^{-i\chi}g_R^{-+}g_L^{+-}) \end{pmatrix}$$

To evaluate this matrix determinant we use the property $g^{++} = g^{--}$, valid in the low energy approximation. Then,

$$\det(\mathbb{I} - \hat{\mathcal{V}}'\hat{g}_R\hat{\mathcal{V}}'^{\dagger}\hat{g}_L) = t.w.\chi + \mathcal{V}'^2\left(e^{i\chi}g_R^{+-}g_L^{-+} + e^{-i\chi}g_R^{-+}g_L^{+-}\right),$$

where $t.w.\chi$ correspond to terms which do not depend on χ.

$$\frac{\det\left(\mathbb{I} - \hat{\mathcal{V}}'(\chi)\hat{g}_R\hat{\mathcal{V}}'^{\dagger}(\chi)\hat{g}_L\right)}{\det\left(\mathbb{I} - \hat{\mathcal{V}}'(0)\hat{g}_R\hat{\mathcal{V}}'^{\dagger}(0)\hat{g}_L\right)} = \frac{t.w.\chi + \mathcal{V}'^2\left(e^{i\chi}g_R^{+-}g_L^{-+} + e^{-i\chi}g_R^{-+}g_L^{+-}\right)}{t.w.\chi + \mathcal{V}'^2\left(g_R^{+-}g_L^{-+} + g_R^{-+}g_L^{+-}\right)}$$

$$= 1 + \mathcal{V}'^2\frac{(e^{i\chi}-1)g_R^{+-}g_L^{-+} + (e^{-i\chi}-1)g_R^{-+}g_L^{+-}}{t.w.\chi + \mathcal{V}'^2\left(g_R^{+-}g_L^{-+} + g_R^{-+}g_L^{+-}\right)}$$

$$\tag{A.86}$$

From now on we need the explicit expressions of the unperturbed GFs. In the low energy approximation $g^{A,R} = \pm i/\mathcal{V}$, then, if n denotes the Fermi function for the corresponding lead we have

$$g^{+-} = n(g^A - g^R) = \frac{2in}{\mathcal{V}} \qquad\qquad g^{-+} = (n-1)(g^A - g^R) = (n-1)\frac{2i}{\mathcal{V}}$$

$$g^{++} = (2n-1)\frac{i}{\mathcal{V}} = g^{--}$$

Using this approximation the denominator $D \equiv t.w.\chi + \mathcal{V}'^2\left(g_R^{+-}g_L^{-+} + g_R^{-+}g_L^{+-}\right)$ in Eq. (A.86) simplifies to

$$D = [1 - \alpha(-1)]^2 + 0 = (1+\alpha)^2, \tag{A.87}$$

where, as in Eq. (9.24), $\alpha = (\mathcal{V}'/\mathcal{V})^2$. Then

$$\frac{\det\left(\mathbb{I} - \hat{\mathcal{V}}'(\chi)\hat{g}_R\hat{\mathcal{V}}'^{\dagger}(\chi)\hat{g}_L\right)}{\det\left(\mathbb{I} - \hat{\mathcal{V}}'(0)\hat{g}_R\hat{\mathcal{V}}'^{\dagger}(0)\hat{g}_L\right)} = 1 + \tau\left[(e^{i\chi}-1)n_R(1-n_L) + (e^{-i\chi}-1)n_L(1-n_R)\right],$$

where τ is the transmission coefficient defined in Eq. (9.24). We finally obtain:

$$S(\chi, t_0) = \frac{t_0}{2\pi} \int d\omega \log\{1 + \tau \left[(e^{i\chi} - 1)n_R(1 - n_L) + (e^{-i\chi} - 1)n_L(1 - n_R)\right]\}$$

$$(A.88)$$

(ii) In order to obtain C_1 and C_2 we derive Eq. (A.88)

$$\frac{\partial S}{\partial \chi} = \frac{t_0}{2\pi} \int d\omega \frac{i\tau \left[e^{i\chi}n_R(1 - n_L) - e^{-i\chi}n_L(1 - n_R)\right]}{1 + \tau \left[(e^{i\chi} - 1)n_R(1 - n_L) + (e^{-i\chi} - 1)n_L(1 - n_R)\right]}$$

$$(A.89)$$

$$\frac{\partial^2 S}{\partial \chi^2} = \frac{t_0}{2\pi} \int d\omega \left\{ -\frac{\left[i\tau \left(e^{i\chi}n_R(1 - n_L) - e^{-i\chi}n_L(1 - n_R)\right)\right]^2}{\left[1 + \tau \left((e^{i\chi} - 1)n_R(1 - n_L) + (e^{-i\chi} - 1)n_L(1 - n_R)\right)\right]^2} \right.$$
$$\left. + \frac{-\tau \left[e^{i\chi}n_R(1 - n_L) + e^{-i\chi}n_L(1 - n_R)\right]}{1 + \tau \left[(e^{i\chi} - 1)n_R(1 - n_L) + (e^{-i\chi} - 1)n_L(1 - n_R)\right]} \right\}$$

$$(A.90)$$

Evaluating these expressions for $\chi = 0$ we obtain

$$\left.\frac{\partial S}{\partial \chi}\right|_{\chi=0} = \frac{i t_0 \tau}{2\pi} \int d\omega \left[n_R(1 - n_L) - n_L(1 - n_R)\right] = \frac{i t_0 \tau}{2\pi} \int d\omega \, (n_R - n_L)$$

$$(A.91)$$

$$\left.\frac{\partial^2 S}{\partial \chi^2}\right|_{\chi=0} = \frac{t_0}{2\pi} \int d\omega \left\{\tau^2 \left[n_R(1 - n_L) - n_L(1 - n_R)\right]^2 - \tau \left[n_R(1 - n_L) + n_L(1 - n_R)\right]\right\}$$

$$(A.92)$$

but $[n_R(1 - n_L) - n_L(1 - n_R)]^2 = (n_R - n_L)^2$. Then

$$\left.\frac{\partial^2 S}{\partial \chi^2}\right|_{\chi=0} = \frac{t_0}{2\pi} \int d\omega \left\{\tau^2(n_R - n_L)^2 - \tau \left[n_R(1 - n_L) + n_L(1 - n_R)\right]\right\} \quad (A.93)$$

We thus obtain

$$C_1 = -i \left.\frac{\partial S}{\partial \chi}\right|_{\chi=0} = \frac{t_0 \tau}{2\pi} \int d\omega \, (n_R - n_L) \qquad (A.94)$$

So that $\langle I \rangle / C_1 = -e/t_0$, and C_1 corresponds to the mean number of particles which are transferred in t_0 in a steady state with a mean current $\langle I \rangle$.

For the second cumulant we have

$$C_2 = -\left.\frac{\partial^2 S}{\partial \chi^2}\right|_{\chi=0} = \frac{t_0}{2\pi} \int d\omega \left\{-\tau^2(n_R - n_L)^2 + \tau \left[n_R(1 - n_L) + n_L(1 - n_R)\right]\right\}$$

$$(A.95)$$

which is clearly related to

$$S_I(0) = \frac{4e^2}{h} \int d\omega' \left\{ \tau \left[n_L(1 - n_R) + n_R(1 - n_L) \right] - \tau^2 (n_R - n_L)^2 \right\}. \quad \text{(A.96)}$$

Exercise II.10: Boundary GF for a BCS lead

As discussed in Sect. 9.3.1, the boundary GF for a semi-infinite chain in Nambu space is given by

$$\hat{g}_S = \left[\omega \hat{\sigma}_0 - \hat{\epsilon} - \hat{v} \hat{g}_S \hat{v} \right]^{-1} \qquad \text{with } \hat{\epsilon} = (\epsilon - \mu)\hat{\sigma}_z + \Delta \hat{\sigma}_x \text{ and } \hat{v} = v \hat{\sigma}_z \quad \text{(A.97)}$$

To simplify we consider $\epsilon - \mu = 0$ (electron-hole symmetry), and propose a solution of the type $\hat{g}_S = g \hat{\sigma}_0 + f \hat{\sigma}_x$. Thus,

$$\left[\omega \hat{\sigma}_0 - \hat{\epsilon} - \hat{v} \hat{g}_S \hat{v} \right]^{-1} = \frac{1}{(\omega - v^2 g)^2 - (\Delta - v^2 f)^2} \begin{pmatrix} \omega - v^2 g & \Delta - v^2 f \\ \Delta - v^2 f & \omega - v^2 g \end{pmatrix}. \quad \text{(A.98)}$$

Then, Eq. (A.97) can be rewritten as

$$\begin{pmatrix} g & f \\ f & g \end{pmatrix} = \frac{1}{(\omega - v^2 g)^2 - (\Delta - v^2 f)^2} \begin{pmatrix} \omega - v^2 g & \Delta - v^2 f \\ \Delta - v^2 f & \omega - v^2 g \end{pmatrix} \quad \text{(A.99)}$$

or, equivalently

$$g = \frac{\omega - v^2 g}{(\omega - v^2 g)^2 - (\Delta - v^2 f)^2} \qquad f = \frac{\Delta - v^2 f}{(\omega - v^2 g)^2 - (\Delta - v^2 f)^2} \quad \text{(A.100)}$$

Dividing both expressions we obtain

$$\frac{\omega - v^2 g}{\Delta - v^2 f} = \frac{g}{f} \iff (\omega - v^2 g) f = (\Delta - v^2 f) g \iff \frac{\omega}{\Delta} f = g$$

and substituting in Eq. (A.100),

$$f = \frac{\Delta - v^2 f}{(\omega - v^2 \frac{\omega}{\Delta} f)^2 - (\Delta - v^2 f)^2} = \frac{\Delta - v^2 f}{\frac{\omega^2}{\Delta^2}(\Delta - v^2 f)^2 - (\Delta - v^2 f)^2}$$

$$= \frac{1}{\Delta - v^2 f} \cdot \frac{1}{\frac{\omega^2}{\Delta^2} - 1}$$

Then

$$f(v^2 f - \Delta) - \frac{\Delta^2}{\Delta^2 - \omega^2} = 0 \iff f = \frac{\Delta \pm \sqrt{\Delta^2 + \frac{4v^2\Delta^2}{\Delta^2 - \omega^2}}}{2v^2} = \frac{\Delta}{2v^2}\left[1 \pm \sqrt{\frac{4v^2 + \Delta^2 - \omega^2}{\Delta^2 - \omega^2}}\right]$$
(A.101)

Therefore,

$$g = \frac{\omega}{2v^2}\left[1 \pm \sqrt{\frac{4v^2 + \Delta^2 - \omega^2}{\Delta^2 - \omega^2}}\right]$$
(A.102)

We select the $(-)$ solution so that $g \sim 1/\omega$ when $\omega \to \pm\infty$. In this way

$$g = \frac{\omega}{2v^2}\left[1 - \sqrt{\frac{4v^2 + \Delta^2 - \omega^2}{\Delta^2 - \omega^2}}\right]$$
(A.103)

$$f = \frac{\Delta}{2v^2}\left[1 - \sqrt{\frac{4v^2 + \Delta^2 - \omega^2}{\Delta^2 - \omega^2}}\right]$$
(A.104)

Exercise II.11: Andreev reflection

(i) In Nambu space, the current operator is defined as

$$\hat{I}_{LR} = \frac{ie}{\hbar}\left(\hat{\Psi}_L^\dagger \hat{\sigma}_z \hat{v}' \hat{\Psi}_R - \hat{\Psi}_R^\dagger \hat{\sigma}_z \hat{v}' \hat{\Psi}_L\right) \qquad \text{with} \qquad \hat{v}' = v'\hat{\sigma}_z = \begin{pmatrix} v' & 0 \\ 0 & -v' \end{pmatrix}$$
(A.105)

We first calculate the following products

$$\hat{\Psi}_L^\dagger \hat{\sigma}_z \hat{v}' \hat{\Psi}_R = v'\begin{pmatrix} c_{L\uparrow}^\dagger & c_{L\downarrow} \end{pmatrix}\begin{pmatrix} c_{R\uparrow} \\ c_{R\downarrow}^\dagger \end{pmatrix} = v'\left(c_{L\uparrow}^\dagger c_{R\uparrow} + c_{L\downarrow} c_{R\downarrow}^\dagger\right) = v'\left(c_{L\uparrow}^\dagger c_{R\uparrow} - c_{R\downarrow}^\dagger c_{L\downarrow}\right)$$

$$\hat{\Psi}_R^\dagger \hat{\sigma}_z \hat{v}' \hat{\Psi}_L = v'\begin{pmatrix} c_{R\uparrow}^\dagger & c_{R\downarrow} \end{pmatrix}\begin{pmatrix} c_{L\uparrow} \\ c_{L\downarrow}^\dagger \end{pmatrix} = v'\left(c_{R\uparrow}^\dagger c_{L\uparrow} + c_{R\downarrow} c_{L\downarrow}^\dagger\right) = v'\left(c_{R\uparrow}^\dagger c_{L\uparrow} - c_{L\downarrow}^\dagger c_{R\downarrow}\right).$$

On the other hand, the G_{LR}^{+-} and G_{RL}^{+-} Green functions are

$$G_{LR}^{+-}(t,t) = i\begin{pmatrix} \langle c_{R\uparrow}^\dagger(t)c_{L\uparrow}(t)\rangle & \langle c_{R\downarrow}^\dagger(t)c_{L\uparrow}(t)\rangle \\ \langle c_{R\uparrow}^\dagger(t)c_{L\downarrow}(t)\rangle & \langle c_{R\downarrow}^\dagger(t)c_{L\downarrow}(t)\rangle \end{pmatrix} \qquad G_{RL}^{+-}(t,t) = i\begin{pmatrix} \langle c_{L\uparrow}^\dagger(t)c_{R\uparrow}(t)\rangle & \langle c_{L\downarrow}^\dagger(t)c_{R\uparrow}(t)\rangle \\ \langle c_{L\uparrow}^\dagger(t)c_{R\downarrow}(t)\rangle & \langle c_{L\downarrow}^\dagger(t)c_{R\downarrow}(t)\rangle \end{pmatrix}.$$

Then

$$\mathrm{Tr}\left[\hat{\sigma}_z \hat{v}' \left(G_{RL}^{+-}(t,t) - G_{LR}^{+-}(t,t)\right)\right] = \mathrm{Tr}\left[v'\left(G_{RL}^{+-}(t,t) - G_{LR}^{+-}(t,t)\right)\right]$$
$$= i\langle\hat{\Psi}_L^\dagger(t)\hat{\sigma}_z\hat{v}'\hat{\Psi}_R(t)\rangle - i\langle\hat{\Psi}_R^\dagger(t)\hat{\sigma}_z\hat{v}'\hat{\Psi}_L(t)\rangle,$$

and,

$$\langle \hat{I}_{LR} \rangle = \frac{ie}{\hbar} \langle \hat{\Psi}_L^\dagger(t)\hat{\sigma}_z \hat{v}' \hat{\Psi}_R(t) - \hat{\Psi}_R^\dagger(t)\hat{\sigma}_z \hat{v}' \hat{\Psi}_L(t) \rangle = \frac{e}{\hbar} \mathrm{Tr}\left[\hat{\sigma}_z \hat{v}' \left(G_{RL}^{+-}(t,t) - G_{LR}^{+-}(t,t) \right) \right] .$$

$$(A.106)$$

In frequency representation one finally obtains

$$\langle \hat{I}_{LR} \rangle = \frac{e}{h} \int d\omega \, \mathrm{Tr}\left[\hat{\sigma}_z \hat{v}' \left(G_{RL}^{+-}(\omega) - G_{LR}^{+-}(\omega) \right) \right]$$

$$(A.107)$$

(ii) Applying the Langreth rules we can write

$$\hat{G}_{RL}^{+-} = (\hat{G}_{RR}\hat{v}'\hat{g}_L)^{+-} = \hat{G}_{RR}^{+-}\hat{v}'\hat{g}_L^A + \hat{G}_{RR}^R\hat{v}'\hat{g}_L^{+-}$$
$$\hat{G}_{LR}^{+-} = (\hat{g}_L\hat{v}'\hat{G}_{RR})^{+-} = \hat{g}_L^{+-}\hat{v}'\hat{G}_{RR}^A + \hat{g}_L^R\hat{v}'\hat{G}_{RR}^{+-} .$$

Substituting in Eq. (A.107) and applying the properties of the trace we obtain

$$\mathrm{Tr}\left[\hat{\sigma}_z \hat{v}' \left(G_{RL}^{+-}(\omega) - G_{LR}^{+-}(\omega) \right) \right] = v' \mathrm{Tr}\left[\hat{G}_{RR}^{-+}\hat{v}'\hat{g}_L^{+-} - \hat{G}_{RR}^{+-}\hat{v}'\hat{g}_L^{-+} \right] \quad (A.108)$$

$$+ v' \mathrm{Tr}\left[\hat{G}_{RR}^{+-}\hat{v}'\hat{g}_L^R + \hat{G}_{RR}^A\hat{v}'\hat{g}_L^{+-} - \hat{G}_{RR}^A\hat{g}_L^{+-}\hat{v}' - \hat{G}_{RR}^{+-}\hat{g}_L^R\hat{v}' \right] \quad (A.109)$$

Finally, taking into account that the L lead is in the normal state ($L \equiv N$) and thus its anomalous components are zero, so that $\hat{g}_L^R$ and $\hat{v}'$ commute, the last term in (A.109) vanishes and we thus get Eq. (9.60), i.e.

$$\langle \hat{I}_{LR} \rangle = \frac{ev'}{h} \int d\omega \, \mathrm{Tr}\left[\hat{G}_{RR}^{-+}\hat{v}'\hat{g}_L^{+-} - \hat{G}_{RR}^{+-}\hat{v}'\hat{g}_L^{-+} \right] .$$

$$(A.110)$$

(iii) At $T = 0$ and $V < \Delta$ the frequency integration in (9.60) is restricted to $|\omega| < \Delta$, where $\hat{g}_R^{+-}(\omega)$ and $\hat{g}^{-+}(\omega)$ vanish. Within this frequency range we thus have

$$\hat{G}_{RR}^{+-,-+} = \hat{G}_{RR}^R\hat{v}'\hat{g}_L^{+-,-+}\hat{v}'\hat{G}_{RR}^A .$$

$$(A.111)$$

When replaced into Eq. (A.107) only terms in $g_e^{+-}g_h^{-+}$ survive in this range. Within the low energy approximation we have $g_e^{+-}(\omega)g_h^{-+}(\omega) = 4/v^2$ for $|\omega| < V$, thus from (9.60)

$$\langle I_{LR} \rangle = \frac{4v'^4}{v^2} \int\limits_{-eV}^{eV} d\omega \left[|(G_{RR}^A)_{12}|^2 + |(G_{RR}^A)_{21}|^2 \right] ,$$

$$(A.112)$$

from which the Andreev conductance expression can be derived.

(iv) From the Dyson equation of $\hat{G}^A_{RR}$

$$\hat{G}^A_{RR} = \hat{g}^A_R \left(\hat{\mathbb{I}} - \hat{v}' \hat{g}^A_L \hat{v}' \hat{g}^A_R \right)^{-1} ,$$

we obtain

$$\left(\hat{G}^A_{RR} \right)_{12} = \frac{1}{D} \left[g^A v'^2 g^A_e f^A + f^A \left(1 - v'^2 g^A_h g^A \right) \right] = \frac{f^A}{D} \left[1 + v'^2 g^A (g^A_e - g^A_h) \right] ,$$

$$(A.113)$$

where $D \equiv \det \left(\hat{\mathbb{I}} - \hat{v}' \hat{g}^A_L \hat{v}' \hat{g}^A_R \right) = \left(1 - v'^2 g^A_e g^A \right) \left(1 - v'^2 g^A_h g^A \right) - v'^4 g^A_e g^A_h (f^A)^2$.

In the low energy approximation

$$g(\omega) = -\frac{\omega}{|v|\sqrt{\Delta^2 - \omega^2}} \qquad\qquad f(\omega) = \frac{\Delta}{|v|\sqrt{\Delta^2 - \omega^2}} \qquad\qquad g^A_e = \frac{i}{|v|} = g^A_h$$

then $g^A(0) = 0$, $f^A(0) = 1/|v|$ y $g^A_e = g^A_h = i/|v|$ and substituting in Eq. (A.113) we obtain

$$\left(\hat{G}^A_{RR} \right)_{12} = \frac{1}{1 + \frac{v'^4}{v^4}} \cdot \frac{1}{|v|}$$

Then, with $\alpha \equiv (v'/v)^2$,

$$R_A = \frac{4v'^4}{v^2} \left| \left(\hat{G}^A_{RR} \right)_{12} \right|^2 = \frac{4v'^4}{v^2} \cdot \frac{1}{\left(1 + \frac{v'^4}{v^4} \right)^2} \cdot \frac{1}{v^2} = \frac{4\alpha^2}{(1 + \alpha^2)^2}$$

On the other hand, as $\tau = 4\alpha/(1 + \alpha)^2$,

$$\frac{\tau^2}{(2 - \tau)^2} = \frac{16\alpha^2}{(1 + \alpha)^4} \cdot \frac{1}{\left[2 - \frac{4\alpha}{(1+\alpha)^2} \right]^2} = \frac{4\alpha^2}{(1 + \alpha^2)^2} = R_A$$

Exercise II.12: Supercurrent and Andreev bound states

(i) We adopt the gauge choice discussed in Sect. 9.3.3, which removes the phase dependence on the leads and assigns it to the hopping matrix, i.e. $\hat{v}' = v'\tau_z e^{i\phi\tau_z/2}$. From Eq. (9.51) for the current and taking into account equilibrium conditions (i.e. $\hat{G}^{+-}_{RL}(\omega) = n_F(\omega) \left[\hat{G}^A_{RL}(\omega) - \hat{G}^R_{RL}(\omega) \right]$) we obtain

$$\langle \hat{I} \rangle = \frac{e}{h} \int d\omega\, n_F(\omega) \mathrm{Tr} \left[\hat{\sigma}_z \left(\hat{v}' \hat{G}^A_{RL}(\omega) + \hat{v}'^\dagger \hat{G}^R_{LR}(\omega) \right) - \hat{\sigma}_z \left(\hat{v}'^\dagger \hat{G}^A_{LR}(\omega) + \hat{v}' \hat{G}^R_{RL}(\omega) \right) \right].$$

$$(A.114)$$

Then, using the general property $\hat{G}^R_{LR}(\omega) = \left[\hat{G}^A_{RL}(\omega)\right]^\dagger$, (A.114) can be written as

$$\langle \hat{I} \rangle = \frac{2e}{h} \int d\omega\, n_F(\omega)\, \text{Re}\left\{ \text{Tr}\left[\hat{\sigma}_z \left(\hat{v}' \hat{G}^A_{RL} - \hat{\sigma}_z \hat{v}'^\dagger \hat{G}^A_{LR} \right) \right] \right\} \tag{A.115}$$

The next step is to show that Eq. (A.115) is equivalent to

$$\langle \hat{I} \rangle = \frac{4e}{h} \int d\omega\, n_F(\omega)\, \text{Re}\left\{ i\frac{\partial}{\partial \phi} \log \det(\mathbb{A}) \right\}, \tag{A.116}$$

with $\mathbb{A} = \hat{\mathbb{I}} - \hat{v}'^\dagger \hat{g}_L \hat{v}' \hat{g}_R$, which can be demonstrated using Jacobi's formula as in Sect. 9.2.

Then, defining $D^A \equiv \det \mathbb{A}$ we find

$$\begin{aligned}
D^A &= 1 - (v')^2 \left[2g^A_R g^A_L - f^A_R f^A_L \left(e^{i\phi} + e^{-i\phi} \right) \right] \\
&\quad + (v')^4 \left[(g^A_R g^A_L)^2 + (f^A_L f^A_R)^2 - (g^A_R f^A_L)^2 - (g^A_L f^A_R)^2 \right].
\end{aligned}$$

and for its ϕ derivative

$$\frac{\partial}{\partial \phi} D^A = (v')^2 f^A_R f^A_L \left(i e^{i\phi} - i e^{-i\phi} \right) = -2(v')^2 f^A_R f^A_L \sin(\phi),$$

so that

$$\frac{\partial}{\partial \phi} \log \det \mathbb{A}(\phi) = -\frac{2(v')^2 f^A_R f^A_L \sin(\phi)}{D^A}$$

Thus,

$$\text{Re}\left\{ i\frac{\partial}{\partial \phi} \log \det(\mathbb{A}) \right\} = \text{Re}\left\{ -2i\, \frac{(v')^2 f^A_R f^A_L \sin(\phi)}{D^A} \right\} = 2(v')^2 \sin(\phi)\, \text{Im}\left\{ \frac{f^A_R f^A_L}{D^A} \right\},$$

which allows to demonstrate that

$$\begin{aligned}
\langle \hat{I} \rangle &= \frac{4e}{h} \int d\omega\, n_F(\omega)\, \text{Re}\left\{ i\frac{\partial}{\partial \phi} \log \det(\mathbb{A}) \right\} \\
&= \frac{8ev'^2}{h} \sin(\phi) \int d\omega\, n_F(\omega)\, \text{Im}\left\{ \frac{f^A_R(\omega) f^A_L(\omega)}{D^A(\omega)} \right\}
\end{aligned} \tag{A.117}$$

(ii) We should first show that $\omega_{1,2} = \pm\Delta\sqrt{1 - \tau \sin^2(\phi/2)}$ correspond to zeros of $D^A(z)$. For that one can start from the expression of $D^A(z)$ in terms of $g_{L,R}$ and $f_{L,R}$,

$$D(z) = 1 - 2(v')^2 \left[g_R g_L - f_R f_L \cos(\phi) \right]$$
$$+ (v')^4 \left[(g_R g_L)^2 + (f_L f_R)^2 - (g_R f_L)^2 - (g_L f_R)^2 \right]$$

Using the identity $g = -\frac{z}{\Delta} f$ for the Nambu components of $\hat{g}$:

$$D(z) = \left\{ 1 + (v')^2 \left[1 - \left(\frac{z}{\Delta} \right)^2 \right] f_R f_L \right\}^2 - 4(v')^2 \sin^2(\phi/2) f_R f_L \qquad \text{(A.118)}$$

Then, assuming a symmetric junction in the low energy approximation we find

$$D(\omega_{1,2}) = \left\{ 1 + v'^2 \tau \sin^2(\phi/2) \frac{1}{v^2 \tau \sin^2(\phi/2)} \right\}^2 - 4v'^2 \sin^2(\phi/2) \frac{1}{v^2 \tau \sin^2(\phi/2)} .$$

where $\tau = 4\alpha/(1+\alpha)^2$, with $\alpha = (v'^2/v)^2$. Then

$$D(\omega_{1,2}) = (1+\alpha)^2 - \frac{4\alpha}{\tau} = 0$$

(iii) At zero temperature $n_F(\omega) \to \theta(-\omega)$, then

$$\langle \hat{I} \rangle = \frac{8ev'^2}{h} \sin(\phi) \int\limits_{-\infty}^{0} d\omega \, \text{Im} \left\{ \frac{f_R^A(\omega) f_L^A(\omega)}{D^A(\omega)} \right\} \qquad \text{(A.119)}$$

In a symmetric junction $f_R^A(\omega) = f_L^A(\omega)$, and its product $f_R^A(\omega) f_L^A(\omega)$ is real for all frequencies. Thus the only contribution to the current comes from the pole at $\omega_2 < 0$. An expansion around ω_2 in the low energy approximation yields

$$D(z) = \frac{8v'^2 \sqrt{1 - \tau \sin^2(\phi/2)}}{v^2 \tau^2 \Delta \sin^2(\phi/2)} (z - \omega_2) + \mathcal{O}(z - \omega_2)^2. \qquad \text{(A.120)}$$

And thus,

$$\text{Im} \left\{ \frac{f_R^A(\omega) f_L^A(\omega)}{D^A(\omega)} \right\} = f_R^A(\omega) f_L^A(\omega) \cdot \frac{\pi}{8} \cdot \frac{v^2 \tau^2 \Delta \sin^2(\phi/2)}{v'^2 \sqrt{1 - \tau \sin^2(\phi/2)}} \cdot \delta(\omega - \omega_2) \qquad \text{(A.121)}$$

which allows to demonstrate that,

$$\langle \hat{I} \rangle = \frac{e\Delta}{2\hbar} \frac{\tau \sin(\phi)}{\sqrt{1 - \tau \sin^2(\phi/2)}} . \qquad \text{(A.122)}$$

Exercise II.13: Transport in Majorana wires

(i) In the low energy approximation we can express the bulk Kitaev GFs (9.71) as

$$\hat{\mathcal{G}}_{00}(\omega) \simeq \frac{-\omega}{\mathcal{V}\sqrt{\Delta^2 - \omega^2}}\sigma_0, \tag{A.123}$$

$$\hat{\mathcal{G}}_{\pm 1,0}(\omega) = \hat{\mathcal{G}}_{0,\mp 1}(\omega) \simeq \frac{\sqrt{\Delta^2 - \omega^2}\sigma_z \mp i\Delta\sigma_y}{\mathcal{V}\sqrt{\Delta^2 - \omega^2}}.$$

Then, we can use the equivalent to Eq. (9.15) but extended to Nambu space, that is

$$\hat{g}^{R,A}(\omega) = \hat{\mathcal{G}}_{11}^{R,A} - \hat{\mathcal{G}}_{10}^{R,A}\left(\hat{\mathcal{G}}_{00}^{R,A}\right)^{-1}\hat{\mathcal{G}}_{01}^{R,A}, \tag{A.124}$$

from which Eq. (9.72) immediately follows.

(ii) As in Sect. 9.3.2 we can use the Hamiltonian approach and obtain the current from an expression like Eq. (9.60). For that we first need to obtain $\hat{G}_{RR}^{+-,-+}$ and $\hat{G}_{RR}^{R,A}$, where now the R lead corresponds to the TS, represented by the Kitaev model which is coupled to the normal L lead. For $\hat{G}_{RR}^{R,A}$ we have

$$\hat{G}_{RR}^{R,A} = \left((\hat{g}^{R,A})^{-1} \pm i\Gamma_N\sigma_0\right)^{-1}, \tag{A.125}$$

where $\Gamma_N = \mathcal{V}'^2/\mathcal{V}$ corresponds to the normal lead self-energy in the low energy approximation. On the other hand, we can obtain $\hat{G}_{RR}^{+-,-+}$ from

$$\hat{G}_{RR}^{+-/-+} = \hat{G}_{RR}^{R}\hat{\Sigma}_L^{+-/-+}\hat{G}_{RR}^{A} + \left(1 + \hat{G}_{RL}^{r}\hat{\mathcal{V}}'\right)\hat{g}^{+-/-+}\left(1 + \hat{\mathcal{V}}'\hat{G}_{LR}^{A}\right), \tag{A.126}$$

where $\hat{g}^{+-} = 2i\,\mathrm{Im}\left(\hat{g}^A\right)f(\omega)$ and $\hat{\Sigma}^{+-} = 2i\Gamma_N\mathrm{diag}\left(f_e, f_h\right)$ with $f_{e,h} = 1/(1 + \exp((\omega \mp \mu_L)/k_BT))$ (the corresponding results for the $-+$ components are obtained by replacing the Fermi factors f and $f_{e,h}$ by $f - 1$ and $f_{e,h} - 1$ respectively.

Substituting these expressions into Eq. (9.60) one can decompose $I = I_A + I_1 + I_2 + I_3$, where

$$I_A = \frac{e}{h}\int d\omega |r_A(\omega)|^2 \left(f_e(\omega) - f_h(\omega)\right), \tag{A.127}$$

where $|r_A|^2 = 4\Gamma^2|G_{RR,12}^A|^2$, is the contribution due to Andreev processes and

$$I_1 = \frac{2e\Gamma}{h}\int d\omega |1 + i\Gamma G_{RR,11}^A|^2\mathrm{Im}g_{11}\left(f_e(\omega) - f_h(\omega)\right)$$

$$I_2 = \frac{4e\Gamma^2}{h}\int d\omega\mathrm{Re}\left[\left(1 - i\Gamma G_{RR,11}^A\right)iG_{RR,21}^A\right]\mathrm{Im}g_{12}\left(f_e(\omega) - f_h(\omega)\right)$$

$$I_3 = \frac{2e\Gamma^3}{h}\int d\omega |G_{RR,12}^A|^2\mathrm{Im}g_{11}\left(f_e(\omega) - f_h(\omega)\right), \tag{A.128}$$

correspond respectively to the normal quasiparticle current, the quasiparticle-pair interference term and the transmission from electron to hole like quasiparticles. By direct calculation one can check that at zero temperature and at subgap voltages the current is only due to Andreev processes. The corresponding contribution to the conductance is

$$G_A(eV < \Delta) = \frac{2e^2}{h} \frac{\tau^2}{4\,(eV/\Delta)^2\,(1 - \tau) + \tau^2}\,. \tag{A.129}$$

The above gap conductance can be obtained from Eqs. (A.128) following a similar procedure.

(iii) The equilibrium mean current in terms of GFs is given by

$$\langle I \rangle = \frac{e\mathcal{V}'}{\hbar} \int \frac{d\omega}{2\pi} f(\omega) 2\mathrm{Re}\left[\left(\hat{G}^A_{RL} - \hat{G}^A_{RL}\right)_{11} e^{-i\phi/2}\right], \tag{A.130}$$

which can be evaluated using the fully dressed GFs (A/R) for the TS-TS junction in the low energy approximation, given by

$$\hat{G}_{RR} = \frac{2\omega}{|\mathcal{V}|\,(1+\alpha)^2\,(\omega^2 - \epsilon_A^2)} \begin{pmatrix} \sqrt{\Delta^2 - \omega^2}\,(1+\alpha) & \Delta\,(\alpha e^{i\phi} - 1) \\ \Delta\,(\alpha e^{-i\phi} - 1) & \sqrt{\Delta^2 - \omega^2}\,(1+\alpha) \end{pmatrix}$$

$$\hat{G}_{LL} = \frac{2\omega}{|\mathcal{V}|\,(1+\alpha)^2\,(\omega^2 - \epsilon_A^2)} \begin{pmatrix} \sqrt{\Delta^2 - \omega^2}\,(1+\alpha) & -\Delta\,(\alpha e^{-i\phi} - 1) \\ -\Delta\,(\alpha e^{i\phi} - 1) & \sqrt{\Delta^2 - \omega^2}\,(1+\alpha) \end{pmatrix}$$

$$\hat{G}_{LR} = \frac{2\mathcal{V}'}{\mathcal{V}^2\,(1+\alpha)^2\,(\omega^2 - \epsilon_A^2)} \begin{pmatrix} 2\Delta^2 \cos\frac{\phi}{2} - \omega^2\,(1+\alpha)\,e^{-i\phi/2} & -2\Delta\sqrt{\Delta^2 - \omega^2}\cos\frac{\phi}{2} \\ 2\Delta\sqrt{\Delta^2 - \omega^2}\cos\frac{\phi}{2} & -2\Delta^2 \cos\frac{\phi}{2} + \omega^2\,(1+\alpha)\,e^{i\phi/2} \end{pmatrix}$$

$$\tag{A.131}$$

where $\epsilon_A = \sqrt{\tau}\Delta \cos(\phi/2)$, which demonstrates that the integrand in (A.130) exhibit poles at $\omega_{1,2} \equiv \pm\epsilon_A(\phi)$.

Part III: Path Integral Formulation

Exercise III.1: Wick's theorem from path integration

The proof of Wick's theorem uses the fact that the a particle time-ordered product can be obtained from differentiation of the generating functional. The present two-particle time-ordered product satisfies:

$$\langle \Psi_1 \Psi_2 \bar{\Psi}_{1'} \Psi_{2'} \rangle = \frac{1}{Z} \frac{\delta^4 Z}{\delta\eta_{2'}\delta\eta_{1'}\delta\bar{\eta}_2\delta\bar{\eta}_1}\Big|_{\eta=\bar{\eta}=0} \tag{A.132}$$

By taking the derivative of the r.h.s of the expression above for $Z[\bar{\eta}, \eta]$ we find:

$$
\frac{\delta Z}{\delta\eta_{2'}\delta\eta_{1'}\delta\bar{\eta}_2}(|A|(\sum_j A_{1j}^{-1}\eta_j)e^{\bar{\eta}A^{-1}\eta})|_{\eta=\bar{\eta}=0}
$$

$$
= \frac{\delta Z}{\delta\eta_{2'}\delta\eta_{1'}}(-|A|(\sum_{k_1} A_{1k_1}^{-1}\eta_{k_1})\sum_{k_2} A_{2k_2}^{-1}\eta_{k_2})e^{\bar{\eta}A^{-1}\eta})|_{\eta=\bar{\eta}=0}
$$

$$
= |A|(-A_{11'}^{-1}A_{22'}^{-1} + A_{12'}^{-1}A_{21'}^{-1}). \tag{A.133}
$$

So we have proved Wick's theorem on the two-particle time-ordered product:

$$
\langle \Psi_1\Psi_2\bar{\Psi}_{1'}\bar{\Psi}_{2'}\rangle = -G_{11'}^{(0)}G_{22'}^{(0)} + G_{12'}^{(0)}G_{21'}^{(0)}. \tag{A.134}
$$

Exercise III.2: Electron-phonon coupling

(i) The partition function for the electron-phonon hamiltonian can be expressed as a path integral:

$$
Z = \int \mathcal{D}\bar{\Psi}\mathcal{D}\Psi \int \mathcal{D}\bar{\Phi}\mathcal{D}\Phi e^{-S[\bar{\Psi},\Psi,\bar{\Phi},\Phi]} \tag{A.135}
$$

where the action reads:

$$
\begin{aligned}
S[\bar{\Psi}, \Psi, \bar{\Phi}, \Phi] = &\sum_{\mathbf{k},n,\sigma} \bar{\Psi}_{\mathbf{k}\sigma}(i\omega_n)(-i\omega_n + \epsilon_{\mathbf{k}})\Psi_{\mathbf{k}\sigma}(i\omega_n) \\
&+ \sum_{\mathbf{q},n,\lambda} \bar{\Phi}_{\mathbf{q}\lambda}(i\nu_n)(-i\nu_n + \omega_{\mathbf{q}\lambda})\Phi_{\mathbf{q}\lambda}(i\nu_n) \\
&+ \sum_{\mathbf{q},n,\lambda} g_{\mathbf{q}\lambda}n_{\mathbf{q}}(i\omega_n)[\Phi_{\mathbf{q}\lambda}(i\omega_n) + \bar{\Phi}_{-\mathbf{q}\lambda}(i\omega_n)].
\end{aligned} \tag{A.136}
$$

where: $n_{\mathbf{q}}(i\omega_n) = \sum_{\mathbf{k},m,\sigma} \bar{\Psi}_{\mathbf{k}+\mathbf{q},\sigma}(i\omega_m + i\nu_n)\Psi_{\mathbf{k},\sigma}(i\omega_m)$.

(ii) We now can integrate over the phonon modes in the partition function through Gaussian integration:

$$
\int \mathcal{D}\bar{\Phi}\mathcal{D}\Phi e^{-\sum_{\mathbf{q},n,\lambda}\bar{\Phi}_{\mathbf{q}\lambda}(i\nu_n)(-i\nu_n+\omega_{\mathbf{q}\lambda})\Phi_{\mathbf{q}\lambda}(i\nu_n)-\sum_{\mathbf{q},n,\lambda}g_{\mathbf{q}\lambda}n_{\mathbf{q}}(i\nu_n)[\Phi_{\mathbf{q}\lambda}(i\nu_n)+\bar{\Phi}_{-\mathbf{q}\lambda}(i\nu_n)]}.
$$

$$
= \frac{1}{Z_{ph}}e^{\sum_{\mathbf{q},n,\lambda}g_{\mathbf{q}\lambda}n_{\mathbf{q}}(i\nu_n)(-i\nu_n+\omega_{\mathbf{q}\lambda})^{-1}g_{-\mathbf{q}\lambda}n_{-\mathbf{q}}(i\nu_n)} \tag{A.137}
$$

where:

$$
Z_{ph} = \prod_{n,\mathbf{q},\lambda}(-i\nu_n + \omega_{\mathbf{q}\lambda}), \tag{A.138}
$$

is the partition function of free phonons which is a trivial constant. By rewriting the effective electron-phonon interaction as:

$$
e^{\sum_{\mathbf{q},n,\lambda}g_{\mathbf{q}\lambda}n_{\mathbf{q}}(i\nu_n)(-i\nu_n+\omega_{\mathbf{q}\lambda})^{-1}g_{-\mathbf{q}\lambda}n_{-\mathbf{q}}(i\nu_n)} = e^{\frac{1}{2}\sum_{\mathbf{q},n,\lambda}g_{\mathbf{q}\lambda}g_{-\mathbf{q}\lambda}n_{\mathbf{q}}(i\nu_n)n_{-\mathbf{q}}(i\nu_n)\frac{2\omega_{\mathbf{q}}}{\nu_n^2+\omega_{\mathbf{q}\lambda}^2}}, \tag{A.139}
$$

the partition function can be expressed as:

$$Z = \frac{1}{Z_{ph}} \int \mathcal{D}\bar{\Psi}\mathcal{D}\Psi \, e^{-S_{eff}[\bar{\Psi},\Psi]} \tag{A.140}$$

with the effective action for the fermions:

$$S_{eff}[\bar{\Psi},\Psi] = \sum_{\mathbf{k},n,\sigma} \bar{\Psi}_{\mathbf{k}\sigma}(i\omega_n)(-i\omega_n + \epsilon_{\mathbf{k}})\Psi_{\mathbf{k}\sigma}(i\omega_n)$$

$$- \frac{1}{2}\sum_{\mathbf{q},n,\lambda} g_{\mathbf{q}\lambda}g_{-\mathbf{q}\lambda}n_{\mathbf{q}}(i\nu_n)\frac{2\omega_{\mathbf{q}\lambda}}{\nu_n^2 + \omega_{\mathbf{q}\lambda}^2}n_{-\mathbf{q}}(i\nu_n), \tag{A.141}$$

analytical continuation $i\omega_n \to \omega + i\eta$ gives an effective attractive interaction between electrons when $\omega < \omega_{\mathbf{q}}$, consistent with our previous result using the finite temperature Green function formalism.

Exercise III.3: Keldysh GFs from the discretized functional integral

As a first step notice that we can write Eq. (10.124) as

$$i\hat{G}^{-1} = \begin{pmatrix} \hat{M}^{++} & \hat{M}^{+-} \\ \hat{M}^{-+} & \hat{M}^{--} \end{pmatrix}, \tag{A.142}$$

where $\hat{M}^{\alpha\beta}$ are $N \times N$ matrices, with $M_{ij}^{++,--} = -\delta_{ij} + h_{\mp}\delta_{i,j+1}$, $M_{ij}^{+-} = \rho\delta_{i1}\delta_{jN}$ y $M_{ij}^{-+} = \delta_{i1}\delta_{jN}$, $h_{\pm} = 1 \pm i\omega_0\delta t$ and ρ such that $\text{Tr}\left[\hat{\rho}(0)\right] = 1/(1-\rho)$. It is then straightforward to show that

$$\left(\hat{M}^{++,--}\right)_{ij}^{-1} = -\theta_{ij}h_{\mp}^{i-j} .$$

We can now use the properties of a block diagonal matrix

$$\begin{pmatrix} A & B \\ C & D \end{pmatrix}^{-1} = \begin{pmatrix} (A - BD^{-1}C)^{-1} & 0 \\ 0 & (D - CA^{-1}B)^{-1} \end{pmatrix}\begin{pmatrix} I & -BD^{-1} \\ -CA^{-1} & I \end{pmatrix}.$$

to get the final result.

Exercise III.4: RG approach to the 1D Hubbard model

(i) This exercise is solved by evaluating all Wick contractions arising from the second order correction to S^0:

$$\langle S_{int}^2[\bar{\Psi}+\bar{\Psi}',\Psi+\Psi']\rangle_0 = \lambda_c\lambda_s\langle\int dx\, J_L(x)J_R(x)\int dy\,\mathbf{J_L}(y)\cdot\mathbf{J_R}(y)\rangle_0. \tag{A.143}$$

which couples the spin and charge sectors. A cancellation of the contractions involving the four fast fields occurs. A similar analysis shows that the second order $\lambda_U\lambda_s$

corrections are also zero. This implies that there spin-charge separation occurs to second order in the coupling λ.

(ii) In this case the factor of 4 in the flow of the λ_U coupling arises from evaluating the Wick contractions of the second order correction:

$$\langle S_{int}^2[\bar{\Psi} + \bar{\slashed{\Psi}}, \Psi + \slashed{\Psi}]\rangle_0 = \lambda_c\lambda_U \langle \int dx\, J_L(x) J_R(x) \int dy((\bar{\Psi}_{L\gamma}(y) + \bar{\slashed{\Psi}}_{L\gamma}(y))$$
$$(\Psi_{R\gamma}(y) + \slashed{\Psi}_{R\gamma}(y)))^2\rangle_0.$$

References

Beenakker, C.W.J.: Three "universal" mesoscopic josephson effects. In: Fukuyama, H., Ando, T. (eds.) Transport Phenomena in Mesoscopic Systems, pp. 235–253. Springer, Berlin (1992)

Blonder, G.E., Tinkham, M., Klapwijk, T.M.: Transition from metallic to tunneling regimes in superconducting microconstrictions: excess current, charge imbalance, and supercurrent conversion. Phys. Rev. B **25**, 4515–4532 (1982)

Büttiker, M.: Scattering theory of thermal and excess noise in open conductors. Phys. Rev. Lett. **65**, 2901–2904 (1990)

Cavalleri, A.: Photo-induced superconductivity. Contemp. Phys. **59**(1), 31–46 (2018)

Cuevas, J.C., Martín-Rodero, A., Yeyati, A.L.: Hamiltonian approach to the transport properties of superconducting quantum point contacts. Phys. Rev. B **54**, 7366–7379 (1996)

Ferrer, J., Martín-Rodero, A., Flores, F.: Contact resistance in the scanning tunneling microscope at very small distances. Phys. Rev. B **38**, 10113–10115 (1988)

Frölich, H.: Electrons in lattice fields. Adv. Phys. **3**(11), 325–361 (1954). https://doi.org/10.1080/00018735400101213

Josephson, B.: Possible new effects in superconductive tunnelling. Phys. Lett. **1**(7), 251–253 (1962)

Kadanoff, L.P., Götze, W., Hamblen, D., Hecht, R., Lewis, E.A.S., Palciauskas, V.V., Rayl, M., Swift, J., Aspnes, D., Kane, J.: Static phenomena near critical points: theory and experiment. Rev. Mod. Phys. **39**, 395–431 (1967)

Keldysh, L.V.: Diagram technique for nonequilibrium processes. Zh. Eksp. Teor. Fiz **47**, 1018 (1964)

Kitaev, A.: Unpaired Majorana fermions in quantum wires, Usp. Fiz. Nauk (Suppl) **44**, 131 (2001)

Kitaev, A.: Anyons in an exactly solved model and beyond. Ann. Phys. **321**, 2–111 (2006)

Kubo, R.: Statistical-mechanical theory of irreversible processes. i. general theory and simple applications to magnetic and conduction problems. J. Phys. Soc. Jpn. **12**, 570–586 (1957)

Kulik, I.O., Omelyanchouk, A.N.: The josephson effect in superconducting constrictions: microscopic theory. J. Phys. Colloques **39**, C6–546–C6–547 (1978)

Landau, L.D.: The theory of a fermi liquid. Soviet Phys. JETP **3**, 920–925 (1957)

Levitov, L., Lesovik, G.: Charge distribution in quantum shot noise. JETP Lett. **58** (1993)

Lieb, E.H., Wu, F.Y.: Absence of mott transition in an exact solution of the short-range, one-band model in one dimension. Phys. Rev. Lett. **20**, 1445–1448 (1968)

Luttinger, J.M.: Analytic properties of single-particle propagators for many-fermion systems. Phys. Rev. **121**, 942–949 (1961)

J. Merino and A. L. Yeyati, *Many-Body Techniques in Condensed Matter Physics*, UNITEXT for Physics, https://doi.org/10.1007/978-3-031-55143-7

Luttinger, J.M.: An exactly soluble model of a many-fermion system. J. Math. Phys **4**, 1154–1162 (1963)

Martín-Rodero, A., García-Vidal, F.J., Levy Yeyati, A.: Microscopic theory of josephson mesoscopic constrictions. Phys. Rev. Lett. **72**, 554–557 (1994)

Matsubara, T.: A new approach to quantum-statistical mechanics. Prog. Theor. Phys. **14**, 351–378 (1955)

Merino, J., Greco, A., Drichko, N., Dressel, M.: Non-fermi liquid behavior in nearly charge ordered layered metals. Phys. Rev. Lett. **96**, 216402 (2006)

Rudner, M.S., Lindner, N.H.: Band structure engineering and non-equilibrium dynamics in floquet topological insulators. Nat. Rev. Phys. **2**(5), 229–244 (2020)

Scalapino, D.J., Loh, E., Hirsch, J.E.: d-wave pairing near a spin-density-wave instability. Phys. Rev. B **34**, 8190–8192 (1986)

Shankar, R.: Renormalization-group approach to interacting fermions. Rev. Mod. Phys. **66**, 129–192 (1994)

Stefanucci, G., van Leeuwen, R.: Nonequilibrium Many-Body Theory of Quantum Systems: A Modern Introduction. Cambridge University Press (2013)

Vecino, E., Martín-Rodero, A., Levy Yeyati, A.: Recursion method for nonhomogeneous superconductors: proximity effect in superconductor-ferromagnet nanostructures. Phys. Rev. B **64**, 184502 (2001)

Wilson, K.G.: The renormalization group: critical phenomena and the kondo problem. Rev. Mod. Phys. **47**, 773–840 (1975)

Wilson, K.G., Kogut, J.: The renormalization group and the expansion. Phys. Rep. **12**, 75–199 (1974)

Zazunov, A., Egger, R., Levy Yeyati, A.: Low-energy theory of transport in majorana wire junctions. Phys. Rev. B **94**, 014502 (2016)

MIX
Papier aus verantwortungsvollen Quellen
Paper from responsible sources
FSC® C105338
FSC
www.fsc.org

If you have any concerns about our products,
you can contact us on
ProductSafety@springernature.com

In case Publisher is established outside the EU,
the EU authorized representative is:
Springer Nature Customer Service Center GmbH
Europaplatz 3, 69115 Heidelberg, Germany

Printed by Libri Plureos GmbH
in Hamburg, Germany